Food protein sources

THE INTERNATIONAL BIOLOGICAL PROGRAMME

The International Biological Programme was established by the International Council of Scientific Unions in 1964 as a counterpart of the International Geophysical Year. The subject of the IBP was defined as 'The Biological Basis of Productivity and Human Welfare', and the reason for its establishment was recognition that the rapidly increasing human population called for a better understanding of the environment as a basis for the rational management of natural resources. This could be achieved only on the basis of scientific knowledge, which in many fields of biology and in many parts of the world was felt to be inadequate. At the same time it was recognized that human activities were creating rapid and comprehensive changes in the environment. Thus, in terms of human welfare, the reason for the IBP lay in its promotion of basic knowledge relevant to the needs of man.

The IBP provided the first occasion on which biologists throughout the world were challenged to work together for a common cause. It involved an integrated and concerted examination of a wide range of problems. The Programme was co-ordinated through a series of seven sections representing the major subject areas of research. Four of these sections were concerned with the study of biological productivity on land, in fresh water, and in the seas, together with the processes of photosynthesis and nitrogen fixation. Three sections were concerned with adaptability of human populations, conservation of ecosystems and the use of biological resources.

After a decade of work, the Programme terminated in June 1974 and this series of volumes brings together, in the form of syntheses, the results of national and international activities.

INTERNATIONAL BIOLOGICAL PROGRAMME 4

Food protein sources

EDITED BY

N. W. Pirie

Rothamsted Experimental Station, Harpenden, UK

with advice from

M. S. Swaminathan

Indian Council of Agriculture Research, New Delhi, India

CAMBRIDGE UNIVERSITY PRESS

CAMBRIDGE

LONDON · NEW YORK · MELBOURNE

Published by the Syndics of the Cambridge University Press
The Pitt Building, Trumpington Street, Cambridge CB2 1RP
Bentley House, 200 Euston Road, London NW1 2DB
32 East 57th Street, New York, NY 10022, USA
296 Beaconsfield Parade, Middle Park, Melbourne 3206, Australia

Library of Congress catalogue card number: 74–12962

ISBN: 0 521 20588 3

First published 1975

Printed in Great Britain
at the University Printing House, Cambridge
(Euan Phillips, University Printer)

Contents

List of contributors *page* xiii
Preface xv

Part I. Sources edible after minimal processing

1 Protein-rich cereal seeds 1
A. K. Kaul
2 Varietal improvement of seed legumes in India 9
L. M. Jeswani
3 Minor food seeds 19
D. A. V. Dendy, Bernice Emmett & O. L. Oke
4 Vegetables 27
F. W. Shepherd
5 The *Spirulina* algae 33
N. W. Pirie
6 Green micro-algae 35
H. Tamiya

Part II. Concentrates made by mechanical extraction

7 Protein products from coconuts 43
D. A. V. Dendy
8 Soybeans: processing and products 47
S. J. Circle & A. K. Smith
9 Rapeseed and other crucifers 65
R. Ohlson & R. Sepp
10 Sunflower, safflower, sesame and castor protein 79
Antoinette A. Betschart, C. K. Lyon & G. O. Kohler
11 Groundnut 105
O. L. Oke, R. H. Smith & A. A. Woodham
12 Broad bean 117
A. Hagberg & J. Sjödin
13 Concentrates by wet and dry processing of cereals 121
R. M. Saunders & G. O. Kohler
14 Leaf protein 133
N. W. Pirie
15 Industrial production of leaf protein in the USA 141
G. O. Kohler & E. M. Bickoff

Part III. Concentrates made by biological conversion

16 Protein from non-domesticated herbivores *page* 147
K. L. Blaxter

17 The use of non-protein nitrogen by ruminants 157
T. R. Preston

18 Non-protein nitrogen in pig nutrition 165
R. Braude

19 The domestic non-ruminant animal as consumer and provider of protein 169
A. A. Woodham

20 The conversion of animal products such as wool and feathers into food 179
F. B. Shorland

21 Increasing the direct consumption of fish 187
G. H. O. Burgess

22 Fungi 201
W. E. Trevelyan

23 Yeasts grown on hydrocarbons 215
C. A. Shacklady

24 Variation in the composition of bacteria and yeast and its significance to single-cell protein production 223
C. L. Cooney & S. R. Tannenbaum

Part IV. The use of novel foods

25 Quality standards, safety and legislation 233
F. Aylward

26 Acceptance of novel foods by the consumer 245
R. P. Devadas

Index 253

Table des matières

Liste des collaborateurs *page* xiii

Préface xv

Iière partie. Sources fournissant un produit comestible après une préparation minimale

1 Graines de céréales riches en protéine 1
A. K. Kaul

2 Amélioration variétale des Légumineuses alimentaires en Inde 9
L. M. Jeswani

3 Graines alimentaires secondaires 19
D. A. V. Dendy, Bernice Emmett & O. L. Oke

4 Légumes 27
F. W. Shepherd

5 L'Algue *Spirulina* 33
N. W. Pirie

6 Algues vertes unicellulaires 35
H. Tamiya

IIième partie. Concentrés obtenus par extraction mécanique

7 Produits protéiniques de la noix de coco 43
D. A. V. Dendy

8 Soja: traitement et produits 47
S. J. Circle & A. K. Smith

9 Graine de colza et autres Crucifères 65
R. Ohlson & R. Sepp

10 Protéines du tournesol, du safran, du sésamé et du ricin 79
Antoinette A. Betschart, C. K. Lyon & G. O. Kohler

11 Arachide 105
O. L. Oke, R. H. Smith & A. A. Woodham

12 Fève 117
A. Hagberg & J. Sjödin

13 Concentrés obtenus par le traitement sec et humide des céréales 121
R. M. Saunders & G. O. Kohler

14 Protéines de feuilles 133
N. W. Pirie

15 Production industrielle de protéines de feuilles aux USA 141
G. O. Kohler & E. M. Bickoff

IIIième partie. Concentrés obtenus par transformation biologique

16 Protéines et herbivores non domestiques *page* 147
K. L. Blaxter

17 L'utilisation de l'azote non protéinique par les ruminants 157
T. R. Preston

18 L'azote non protéinique dans l'alimentation du porc 165
R. Braude

19 L'animal domestique non-ruminant en tant que consommateur et producteur de protéines 169
A. A. Woodham

20 La transformation en aliments de produits d'origine animale, tels que la laine et les plumes 179
F. B. Shorland

21 L'accroissement de la consommation du poisson 187
G. H. O. Burgess

22 Champignons 201
W. E. Trevelyan

23 Levures cultivées sur des hydrocarbures 215
C. A. Shacklady

24 Variations de la composition des bactéries et des levures; leur signification pour la production de protéines d'origine unicellulaire 223
C. L. Cooney & S. R. Tannenbaum

IVième partie. L'utilisation d'aliments nouveaux

25 Standards de qualité, controles sanitaires et législation 233
F. Aylward

26 L'acceptation de produits alimentaires nouveaux par le consommateur 245
R. P. Devadas

Index 253

Содержание

Список авторов *страница* xiii

Предисловие xv

Часть I. Ресурсы белка, пригодные в пищу после минимальной обработки

1 Богатые белком семена злаков 1
A. K. Kaul

2 Улучшение разновидностей семенных бобовых в Индии 9
L. M. Jeswani

3 Мелкие пищевые семена 19
D. A. V. Dendy, Bernice Emmett & O. L. Oke

4 Овощи 27
F. W. Shepherd

5 Водоросли *Spirulina* 33
N. W. Pirie

6 Зеленые микроскопические водоросли 35
H. Tamiya

Часть II. Концентраты приготовленные механическим экстрактированием

7 Белковые продукты из косовых орехов 43
D. A. V. Dendy

8 Соевые бобы: обработка и продукты 47
S. J. Circle & A. K. Smith

9 Семена рапса и другие крестоцветные 65
R. Ohlson & R. Sepp

10 Белок подсолнечника, сафлора, сезама и кастора 79
Antoinette A. Betschart, C. K. Lyon & G. O. Kohler

11 Земляной орех 105
O. L. Oke, R. H. Smith & A. A. Woodham

12 Конские бобы 117
A. Hagberg & J. Sjödin

13 Концентрация сырой и сухойобработкой злаков 121
R. M. Saunders & G. O. Kohler

14 Белки листьев 133
N. W. Pirie

15 Промышленное производство белка листьев в США 141
G. O. Kohler & E. M. Bickoff

Часть III. Концентрирование посредством биологического превращения

16 Белки от неодомашненных животных *страница* 147
K. L. Blaxter

17 Потребление небелкового азота жвачными 157
T. R. Preston

18 Небелковый азот в питании свиней 165
R. Braude

19 Домашние нежвачные животные как потребители и производители белка 169
A. A. Woodham

20 Превращение шерсти и перьев животных в пищу 179
F. B. Shorland

21 Увеличение прямого потребления рыбы 187
G. H. O. Burgess

22 Грибы 201
W. E. Trevelyan

23 Выращивание дрожжей на углеводах 215
C. A. Shacklady

24 Вариации в составе бактерий и дрожжей и их значение при продуцировании белка одноклеточными 223
C. L. Cooney & S. R. Tannenbaum

Часть IV. Потребление новых видов пищи

25 Стандарты качества, безопасность и законодательство 233
F. Aylward

26 Приемлемость новых видов пищи для потребителя 245
R. P. Devadas

Указатель 253

Contenido

Colaboradores *página* xiii

Prefacio xv

Parte I. Materiales comestibles tras una manufactura mínima

1 Semillas de cereales ricas en proteína 1
A. K. Kaul

2 Mejora de variedades de semillas leguminosas en la India 9
L. M. Jeswani

3 Semillas alimenticias secundarias 19
D. A. V. Dendy, Bernice Emmett y O. L. Oke

4 Verduras 27
F. W. Shepherd

5 El alga *Spirulina* 33
N. W. Pirie

6 Algas verdes microscópicas 35
H. Tamiya

Parte II. Concentrados obtenidos por extracción mecánica

7 Derivados proteicos del coco 43
D. A. V. Dendy

8 Soja: manufactura y productos 47
S. J. Circle y A. K. Smith

9 Semillas de colza y de otras crucíferas 65
R. Ohlson y R. Sepp

10 Proteínas del girasol, alazor, sésamo y ricino 79
Antoinette A. Betschart, C. K. Lyon y G. O. Kohlet

11 Cacahuete 105
O. L. Oke, R. H. Smith y A. A. Woodham

12 Haba 117
A. Hagberg y J. Sjödin

13 Concentrados por tratamiento húmedo y seco de cereales 121
R. M. Saunders y G. O. Kohler

14 Proteína foliar 133
N. W. Pirie

15 Producción industrial de proteína foliar en EE.UU 141
G. O. Kohler y E. M. Bickoff

Parte III. Concentrados obtenidos por transformación biológica

16 Proteína obtenida de herbívoros silvestres *página* 147
K. L. Blaxter

17 Utilización del nitrógeno no proteico por los rumiantes 157
T. R. Preston

18 El nitrógeno no proteico en la nutrición del cerdo 165
R. Braude

19 El animal doméstico no rumiante como fuente y consumidor de proteína 169
A. A. Woodham

20 Conversión de productos animales, como lana o plumas, en alimentos 179
F. B. Shorland

21 Intensificación del consumo directo de pescado 187
G. H. O. Burgess

22 Hongos 201
W. E. Trevelyan

23 Levaduras que crecen sobre carbohidratos 215
C. A. Shacklady

24 Variación en la composición de bacterias y levaduras, y su importancia en la producción proteica de la célula aislada 223
C. L. Cooney y S. R. Tannenbaum

Parte IV. Empleo de nuevos alimentos

25 Exigencias de calidad, seguridad y legislación 233
F. Aylward

26 Aceptación de nuevos alimentos por el consumidor 245
R. P. Devadas

Indice 253

Contributors

Aylward, F.	Department of Food Science, University of Reading, Reading RG1 5AQ, UK
Betschart, A. A.	Field Crops Laboratory, US Department of Agriculture, Berkeley, California 94710, USA
Bickoff, E. M.	Field Crops Laboratory, US Department of Agriculture, Berkeley, California 94710, USA
Blaxter, K. L.	Rowett Research Institute, Bucksburn, Aberdeen AB2 9SB, UK
Braude, R.	National Institute for Research into Dairying, Reading RG2 9AT, UK
Burgess, G. H. O.	Torry Research Station, P.O. Box 31, Abbey Road, Aberdeen AB9 8DG, UK
Circle, S. J.	Anderson Clayton Foods, W. L. Clayton Research Center, 3333 Central Expressway, Richardson, Texas 75080, USA
Cooney, C. L.	Department of Nutrition and Food Science, Massachusetts Institute of Technology, Cambridge, Mass. 02139, USA
Dendy, D. A. V.	Tropical Products Institute, 56 Gray's Inn Road, London WC1X 8LU, UK
Devadas, R. P.	Sri Avinashilingam Home Science College, Coimbatore 11, India
Emmett, B.	Tropical Products Institute, 56 Gray's Inn Road, London WC1X 8LU, UK
Hagberg, A.	Swedish Seed Association, S-268 00 Savlöv, Sweden
Jeswani, L. M.	Directorate of Pulses Development, C 60 E Park, Mahanager Extension, Lucknow 6, India
Kaul, A. K.	Institut für Strahlenbotanik, Herrenhäuserstrasse 2, 3000 Hannover, West Germany
Kohler, G. O.	Field Crops Laboratory, US Department of Agriculture, Berkeley, California 94710, USA
Lyon, C. K.	Field Crops Laboratory, US Department of Agriculture, Berkeley, California 94710, USA
Ohlson, R.	Alfa-Laval AB, Karlshamns Oljefabriker AB, 292 00 Karlshamn, Sweden

Oke, O. L.	Chemistry Department, University of Ife, Ile-Ife, Nigeria
Pirie, N. W.	Rothamsted Experimental Station, Harpenden, Herts AL5 2JQ, UK
Preston, T. R.	Animal Production and Health Division, FAO, 00100 Rome, Italy*
Saunders, R. M.	Field Crops Laboratory, US Department of Agriculture, Berkeley, California 94710, USA
Sepp, R.	Alfa-Laval AB, Karlshamns Oljefabriker AB, 292 00 Karlshamn, Sweden
Shacklady, C. A.	BP Proteins Ltd., Britannic House, Moor Lane, London EC2Y 9BU, UK
Shepherd, F. W.	Bosbigal, Old Carnon Hill, Carnon Downs, Truro, Cornwall, TR3 6LF, UK
Shorland, F. B.	Department of Food Science and Human Nutrition, Michigan State University, East Lansing, Michigan 48824, USA
Sjödin, J.	Swedish Seed Association, S-268 00 Svalöv, Sweden
Smith, A. K.	Anderson Clayton Foods, W. L. Clayton Research Center, 3333 Central Expressway, Richardson, Texas 75080, USA
Smith, R. H.	Rowett Research Institute, Bucksburn, Aberdeen AB2 9SB, UK
Swaminatham, M. S.	Indian Council of Agricultural Research, Krishi Bhavan, Dr Rajendra Prasad Road, New Delhi, India
Tamiya, H.	Department of Botany, University of Tokyo, Tokyo 113, Japan
Tannenbaum, S. R.	Department of Nutrition and Food Science, Massachusetts Institute of Technology, Cambridge, Mass. 02139, USA
Trevelyan, W. E.	Tropical Products Institute, 56 Gray's Inn Road, London WC1X 8LU, UK
Woodham, A. A.	Rowett Research Institute, Bucksburn, Aberdeen AB2 9SB, UK

* Now in the División Nutrición Ganadera, Comisión Nacional de la Industria Azucarera, Humboldt 56, 2° piso, Mexico 1, Mexico.

Preface

The inclusion of the words 'For Human Welfare' in the full title of IBP led many early enthusiasts for the Programme to assume that the practical application of research would be so widely recognised that there would be no need to deal with it specifically. However, as research proposals began to come in, it became clear that most participants had an academic approach. There was at first some opposition to the idea of having a Section UM (Use and Management): it was set up later than the other sections and several countries did not form national UM committees. Those countries that had UM committees usually attached more importance to the themes 'Plant Gene Pools' and 'Biological Control' than to the other five themes included in the section. Consequently, a synthesis volume on Theme 3, 'Development of Biological Resources', is more a record of the type of work it was hoped would be included in IBP than of work that actually was included.

The nature and mode of operation of IBP were at first widely misunderstood. Some thought it was a sinister conspiracy out to dictate the world's biological research policy, others thought it was a source of funds. Both were wrong. It was a piece of machinery that was often helpful in getting research funds. But the machinery had to be set in motion by whoever wanted the funds. Differences in the time taken by different people to realise that they had a new ally in fund-getting largely explain the uneven distribution of research effort within IBP. Sometimes an existing national or international research structure helped IBP by providing an organisation within or alongside which it would work. Sometimes a pre-existing organisation inhibited IBP activity because nothing more seemed to be needed. This distinction is clear in national programs: some are mainly new work that might not have started without IBP while others list many projects on which active research had been financed for years.

National UM programs demonstrated widespread interest in protein sources. This limitation of the meaning to be given to 'Biological Resources' was congenial to the co-ordinators of Theme 3 because it seemed that more research would be needed to meet the world's present and impending protein need, than to meet the equally important need for dietary components such as energy and vitamins. Methods for meeting the latter needs are known though not fully applied.

The IBP organised a Working Group meeting in Warsaw in 1966 on

'Novel Protein Sources', a symposium in Varna in 1968 which included a section on 'Subsidiary Sources of Protein Food', and a symposium in Stockholm in 1968 on 'Evaluation of Novel Protein Products'. Summaries of the first two were published in *IBP News*, 7 and 12; the complete text of the third was published (Bender *et al.*, 1970). A Technical Group meeting in Coimbatore in 1970 was limited to leaf protein and IBP Handbook 20 (Pirie, 1971) resulted from it.

The published programs contain many projects connected with protein supplies. Unfortunately, some did not get fully organised. A book limited to those that actually started would have covered the subject inadequately. A meeting convened at the time of the International Nutrition Congress in Mexico in 1972 decided to put no such limit on either subject or authorship. It should, however, be recognised that IBP was actively involved in research on protein in algae, coconuts, deer and other wild herbivores, groundnuts, leaves, and seed legumes.

The book is divided into four sections. The divisions between the first three depend on the extent to which the original products of photosynthesis, on which all the foods eaten at present depend, have been altered in the course of making the ultimate food. The last section deals with restraints on the introduction of new foods and the methods that can be used to popularise them.

Many plants, or parts of plants, can be eaten with no more treatment than can be managed in the kitchen. Obviously, some of these protein-rich foods, e.g., cereal seeds, usually get this pre-treatment in a factory. But it could be managed in the kitchen, and it could be dispensed with. Because a palatable diet usually contains some fat and sugar, if it is finally to contain 8 to 12 % of protein in the dry matter, much of it must consist of material containing significantly more protein than this. Arbitrarily, the lower limit for a 'protein source' was set at 15 % protein. That limitation excludes all the underground parts of plants. It also excludes the cereals that supply more than half the protein that is now eaten, but not some of the newer cereals; there is therefore a chapter on them. Some minor seeds are included, because they are interesting and liable to be overlooked, although they contain less than 15 % protein. Blue-green algae are eaten now in parts of Africa, and were at one time eaten in Mexico, having been collected by hand from natural growth in lakes. They are, therefore, properly included in this section even though industrial production is now proposed. By contrast, a chapter in this book argues that green algae, after a phase of popularity in research institutes, are not to be taken seriously as a bulk human food. They are included for historical reasons, because of their role as condiments and pharmaceutical agents, and because a mixture of algae and bacteria, grown on sewage, is used as an animal feed.

Most of the protein sources that are grouped together in the section on concentrates made by mechanical extraction, are regularly eaten in small quantities in some places without extraction. However, it would be unwise to rely on them for a large fraction of the daily protein supply because of the presence of fibre or of components of varying degrees of toxicity. During the extraction, these deleterious components are removed. This section is intended to be illustrative rather than comprehensive. Work on coconut, groundnut, and leaf protein is included because these were the subjects of IBP projects. Work on leaf protein was indeed undertaken on a more international scale than work on any other topic in Theme 3: useful contributions came from India, New Zealand, Nigeria, Sweden and the UK. Choice among the other sources that might have been considered was more arbitrary. Soybeans are included because of the interesting contrast between the sophisticated methods used in industrialised countries, and the traditional methods in which the extracted protein is coagulated with gypsum, or in which deleterious components are destroyed by fermentation. This chapter is designed to stress the idea that each approach has its merits, and that more attention in research institutes could with advantage be given to improving the traditional methods. Rape is included because of its growing importance. It could be argued that cottonseed should be included for the same reason, and it has the added merit of being a by-product. But the gossypol, which makes cottonseed unsuitable as human food in its unprocessed form, is so conveniently concentrated in glands that can be separated from the meal, that no principle not covered by the other examples of seed-fractionation seems to be involved.

Nothing need be lost during mechanical fractionation. A protein source that could not be eaten in its original state is separated into a concentrate that can be eaten, and a remainder that can still be used as animal feed or in other ways. Biological conversion inevitably involves loss because of the metabolism and energy requirement of the converter. Domestic ruminants are the most obvious and important source of protein made by conversion. A chapter on them is not included because the subject is so vast that it would need a whole book to itself – and there are many such books. There are, however, four chapters on mammals that are now widely eaten, and one on mammalian products that could be eaten though they are not eaten at present. Wild animals are of special interest to IBP; it has been criticised for paying too much attention to conservation for its own sake. The choice of species that could have been considered is enormous. Many will regret that there is nothing here on the capybara, kangaroo, or manatee. The last omission is particularly sad because of the role it could play as a source of meat and as a controller of unwanted aquatic vegetation. Instead, the chapter on wild animals is limited to some species that are already being exploited commercially – there had to be omissions to ensure that the book would be of manageable size.

The decision to limit Theme 3, and this book, to protein sources was not taken because proteins have a unique position in human and animal nutrition that is not shared with vitamins and energy sources, it was taken because more problems can be foreseen in meeting human and animal protein needs than in meeting other nutritional needs. It has been known since 1891 that ruminants, by virtue of the synthetic capacities of the rumen flora, can use nitrogen (N) in forms other than the amino acids. This observation has remained curiously underexploited as a means of diminishing the demand for protein in animal feeds but, as one chapter shows, it is now getting purposeful attention. Mammals without a rumen, presumably because of their different, and smaller, microbial population, are less able to use non-protein N. Nevertheless, the ability of non-ruminants to use simple N compounds is the subject of one chapter. The food requirements of non-ruminants are so similar to our own that much of what they usually eat could have been used as human food. So long as the domestic non-ruminants in a country can be maintained on scraps and by foraging, they are admirable; in a country with a food shortage the advantage of keeping so many that they compete with people must be considered carefully. The case for them would be strengthened if their needs could be partly met by simple N compounds. Non-ruminants are more fecund than ruminants and can therefore respond more quickly to changes in the food supply. The case against both types of animal, when kept on potentially arable land or on fodder grown on arable land, is that they return only 10 to 30 % of the protein they eat in forms usually considered edible by people. It may be that we are too prejudiced. One chapter suggests that such animal products as hair and feathers could be made acceptable. If this were done, the effective efficiency of some animal converters would be greatly increased.

Fish and mussels are the concern of other sections of IBP and are the subjects of other books in this series. During the past decade a widespread impression has been fostered that the future of fish lies in its conversion to the tasteless impalpable powders that will not be noticed when added to conventional foods. Many of the protein sources discussed in this book have that form; silence could have suggested agreement with the popular view. There is, therefore, a chapter arguing that there is an unsatisfied market for fish eaten as fish. Without any increase in the amount caught, much would be gained if more of the catch were eaten instead of being turned into animal fodder.

The enzymes produced by the digestive systems of vertebrates are, to a large extent similar. Consequently, no vertebrate is able, unaided, to use a large group of naturally occurring carbon compounds. It is important to remember that, since the development of methods for the industrial fixation of N, protein production is as often restricted by the supply of precursors containing carbon in the reduced form as it is by the supply of compounds containing N. Plants that photosynthesise can use fully oxidised carbon (CO_2): other organisms, living in the atmosphere made by photosynthesis, have perforce to depend upon carbon in a partly reduced state. A more extensive group of such compounds can be exploited by micro-organisms than by animals because the former are equipped with a more varied set of hydrolytic enzymes. Micro-organisms, even when cultivated on substances that vertebrates can metabolise, have a useful role because of their ability to depend entirely on simple compounds for the N needed in protein synthesis.

Most research is either done in the industrialised countries, or is influenced by their needs and academic interests. This is clearly shown by the dominant direction of research on the production of yeast for use as feed or fodder. The favoured substrate is petroleum although that is an exhaustable asset, and although the techniques needed when it is the substrate are more complex than those needed with molasses and other by-products of agriculture. Those who think that protein sources become more interesting and important in proportion to the likelihood that they could be produced from local products for local use, will study the three articles on microbial protein critically. There is much in them that is of universal applicability; some of the techniques could be used by highly skilled people only.

Pessimists, looking superficially at the history of many attempts to introduce new types of food devised by scientists during the past twenty-five years, are apt to conclude that the outlook is gloomy and that people, even, or perhaps especially, when malnourished, are very conservative in their eating habits. Optimists observe that during the same period there have been great changes in the types of commercially promoted food eaten, and increased substitution of processed for traditional foods. A change that is particularly relevant in the context of protein supplies is the catastrophic replacement of breast-feeding by the bottle. Scientists agree with the old dictum: 'A food has no nutritional value if it is not eaten'. Some of them conclude that the poor acceptance of their new foods may be the fault of their techniques of presentation and advocacy, rather than of intrinsic defects in the foods themselves. Two chapters deal with this point. One is on the safety and quality standards that have to be met to satisfy local and international regulations; the other is on the more subtle question of how to interest consumers in a new product, and how to retain interest once it has been aroused. Within a community, just as within a family, no factor is more important than example.

The letter inviting contributions said that the book would cover every major type of protein source and that it was '...intended primarily as a source of information either on the weight of edible protein that can be produced from a hectare within a year, or on the amount that can be made by extraction or conversion from some inedible material. An indication of the nutritive value of the end product should therefore be given'. The subjects treated have the scope envisaged in the original plan, but authors have not always followed the other suggestions. However, the information given should be sufficient to enable the applicability of each method in any country to be assessed shrewdly. No attempt has been made to get uniformity in the period covered by the different chapters. April 1973 was the original date agreed for the receipt of typescripts; a few were received by that date or soon after. As other editors of collective works have found, plans were then seriously upset by a few dilatory authors.

References

Bender, A. E., Kihlberg, R., Löfqvist, B. & Munck, L. (1970). *Evaluation of novel protein products*. Pergamon Press, Oxford.

Pirie, N. W. (ed.) (1971). *Leaf protein: its agronomy, preparation, quality and use*. IBP Handbook 20. Blackwell Scientific Publications, Oxford.

April, 1974 N. W. PIRIE

PART I

Sources edible after minimal processing

1. Protein-rich cereal seeds

A. K. KAUL

At present, cereals supply more than half the protein eaten by people, and in the coming decades they will probably remain the dominant food in the malnourished and undernourished two-thirds of the world. They are equally important as animal feed in countries in which livestock is produced. Apart from their wide agricultural adaptation and the traditional dual role that they have played in sustaining the subsistence farmer and his cattle (both milch and draught), cereals have undergone centuries of evolution in their components to develop the familiar bland taste, reasonable bulkiness, ease of cooking, lack of any anti-nutritional factors, good keeping quality, sufficient blend of minerals and vitamins and above all a favourable NDpCal (net dietary-protein calories) of over 5 %. The genetic enhancement in the quantitative and qualitative properties of cereal proteins is therefore relevant to both short and long term goals of bridging the protein-calorie gap.

During the past seven years there has been a major breakthrough in cereal production through what is commonly referred to as the 'Green Revolution'. This increase in productivity has been attained by the exploitation of the potential of new high-yielding varieties to better agronomic inputs. Almost simultaneously with this achievement, plant breeders started searching for genes which could improve the quantitative and qualitative aspects of grain proteins and thereby the biological value of the new high-yielding strains. It is satisfying to record that already a large number of nutritionally superior genotypes has been isolated from the world germplasm collections of maize, barley, rice, wheat and recently sorghum. This systematic search has now been extended to legumes and minor millets as well. A parallel effort is also under way to artificially induce such positive variation, using various chemical and physical mutagens. Notable results have been obtained in barley, rice, wheat and peas. One of the side effects of this activity is the development of numerous analytical techniques that have made it possible to screen very large populations of plant material for protein quality and biological value. Some of the major achievements are summarised here.

The ultimate goal of a plant breeding programme, aimed at improving the nutritive value of crops, lies in the accumulation of desirable genes

Table 1.1. *Increasing protein productivity through intensive cropping* (Comparison of 1958 and 1968 cropping patterns at the Indian Agricultural Research Institute, New Delhi)

	1958; two-crop rotation				
Yield (kg/ha)		Maize 2000		Wheat 2500	Total 4500
Protein (kg/ha)		160		250	410
Protein as % of yield		8		10	
	1968; four-crop rotation				
	Moong bean	Maize	Potato	Wheat	Total
Yield (kg/ha)	1000	4200	20600	4700	30500
Protein (kg/ha)	200	504	412	752	1868
Protein as % of yield	20	12	2	16	

After Kaul (1968).

in the high-yielding cultivars to enhance the quantity of nutritionally available amino acids and thereby the biological value of the protein. The various approaches are discussed below.

Total protein productivity per unit area can be significantly enhanced through various agronomic manipulations such as foliar applications of nitrogenous fertilizers just before anthesis (Swaminathan *et al.*, 1969) or through multiple and relay cropping of relatively early-maturing genotypes (Anon., 1973 and Table 1.1). The reduction in the maturity period in some crops has made it possible to harvest two crops in areas where only one was possible before. There are vast areas, particularly in the temperate zones, where winter crops could thus be successfully alternated with some summer ones. One need not emphasise the other benefits of incorporating legumes in such intensive cropping patterns. Relay cropping can be immediately practised on at least 8×10^5 ha in India and in future extended onto 32×10^5 ha.

It has been suggested that in the crops that have superior aminogram patterns, such as rice (*Oryza sativa*) and oats (*Avena sativa*), efforts should be concentrated on increasing the total harvest (Table 1.2). As has already been mentioned, this is being achieved through the popularisation of the high-yielding varieties. Yields of up to 11000 kg/ha (= 1100 kg of crude protein or 36 kg of lysine/ha), have been reported in rice. Here it may be pointed out that higher nitrogen inputs in rice have resulted in almost linear increase in the grain protein content (Swaminathan *et al.*, 1969). It has sometimes been argued that the

Table 1.2. *Some essential amino acids (EAA), true digestibility (TD), biological value (BV), net protein utilisation (NPU) and utilisable nitrogen (UN) of some new varieties and old cultivars of important cereals*

Crop	Barley		Maize		Rice		Sorghum	Rye	Oats
	Normal	Hiproly	Normal	Opaque-2	Normal	High Protein	Normal	Normal	Normal
EAA (g/16 g N)									
Lysine	3.69	4.08	2.74	4.48	3.49	2.90	1.83	3.67	4.03
Methionine	1.82	1.99	1.88	1.78	2.07	1.93	1.72	1.68	1.77
Threonine	3.60	3.41	3.31	3.46	3.25	3.31	3.61	3.31	3.63
Isoleucine	3.68	3.77	2.80	3.02	4.46	4.40	4.50	3.11	3.98
N % in DM	1.62	2.99	2.11	1.97	1.29	2.74	2.06	1.46	1.72
Protein value (as %)									
TD	82	85	96	94	101	100	85	77	84
BV	72	76	60	74	65	86	52	78	70
NPU	59	65	57	70	66	86	44	59	59
UN	0.95	1.94	1.21	1.38	—	—	0.99	0.86	1.02

After Eggum (1973).

proteins increased through such agronomic manipulations are of low biological value. However, in human adults where nitrogen balance, rather than growth, is of primary interest this argument may not hold (Kies *et al.*, 1967).

Possibilities of increasing the crude protein content through conventional and mutation breeding methods have been demonstrated in many cereals (Table 1.3). In some cases the increased protein content in the grain has had adverse effects on the yielding capacity and the quality of the proteins. These two limitations need not however be taken as a generalisation since it has been demonstrated that the negative correlation between yield and the essential amino acids, such as lysine, ceases to operate at the higher levels of protein (> 20 %) and that yielding capacity and protein productivity need not always be inversely correlated (Johnson *et al.*, 1970). Rice and oats have a natural safeguard against these adverse effects as they contain little prolamin. (Prolamins are an alcohol soluble fraction, predominant in maize and sorghum, that have poor biological value. They are the first to increase when protein levels of cereals are raised through genetic and agronomic means.) It is presumed that these two crops already possess genes to suppress the synthesis of prolamin. Considerable positive variation has been recorded in the following crop species: *Eleusine coracana*, *Panicum miliare*,

Table 1.3. *Some nutritionally superior genotypes reported in various cereals*

Crop	Highest protein (%)	Highest lysine (%) (g/16 g N)	Remarks
MAIZE (*Zea mays*)	Normal 10	Normal 2.3	
Inbreds	Up to 24	—	Illinois selections, only of theoretical interest.
Opaque-2 (Op-2)	11	4.0	Superior to normal and Fl-2.
Floury-2 (Fl-2)	11	3.5	Both mutants have yet to appear in the commercial varieties.
Opaque-7 (O_7)	Similar to above		
WHEAT (*Triticum aestivum*)	Normal 10	Normal 2.5	
Atlas-66	19.9	3.0	These three genotypes are presently being incorporated in the commercial varieties at Lincoln, Nebraska, USA.
Napal	20.5	3.2	
Atlas × Napal	Up to 24	Up to 3.2	
Yugoslavian mutants	Up to 20	—	Artificially induced mutants.
Indian mutant	Up to 16	3.0	Artificially induced in the high-yielding Mexican variety and released commercially as 'Sharbati Sonora'. High BV, NPU.
BARLEY (*Hordeum vulgare*)	Normal 10	Normal 3.4	
Hiproly	19.59	4.1	Both mutants are presently being backcrossed into commercial varieties.
Hily	16.84	4.2	
'Notch' mutant (India)	20.0	4.0	All the three mutants lack the normal yield potential.
Denmark mutants	13–15	3.5	Likely to appear commercially.
RICE (*Oryza sativa*)	Normal 7	Normal 3.5	
Japanese mutant	Up to 16	—	Induced mutant of theoretical interest only.
Indian mutant	12	—	
IRRI selections	Up to 15	—	
Indian selection (Assam)	14	4.5	Assam, being one of the centres of origin for rice, offers valuable germplasm to breeders.
OATS (*Avena sativa*)	Normal 12	Normal 4.1	
Garland (USA)	17.2	4.1	
Canadian selection	28.5	3.7	Ideal for making protein concentrates.
Israel (*A. sterilis*)	15–30	—	
SORGHUM (*Sorghum vulgare*)	Normal 10	Normal 1.8	
Purdue selections	Up to 26	Up to 3.8	Recently two Opaque-like mutants have been reported.
Indian selections	Up to 20	Up to 3.0	
PEARL MILLET (*Pennisetum typhoides*)	Normal 9	Normal 2.0	
Indian selections	Up to 18	Up to 3.8	

Panicum miliaceum, *Paspalum scrobiculatum*, *Setaria italica*, *Echinochloa colona* (Swaminathan *et al.*, 1970).

Increase in the relative quantity of one or more amino acids in storage proteins, with or without simultaneous increase in the crude protein content, has been demonstrated in maize (*Zea mays*) and barley (*Hordeum vulgare*). In maize the Opaque-2 and Floury-2 genes double the content of lysine and tryptophan; this is ascribed to a reduced amount of zein (maize prolamin) and an altered grain microstructure. The superior aminogram pattern of Opaque-2 resulted in its significantly increased biological value in experiments conducted on rats, pigs, children and adults (Nelson, 1969). In 1968, Swedish workers identified a similar mutant in the Ethiopian collection of barley (Hagberg & Karlsson, 1969). They named it 'Hiproly' due to its high protein and lysine content. The amino acid profiles of these mutants are compared in Table 1.4. Identical high lysine mutants in barley have since been artificially induced by Doll (1972) in Denmark and by Bansal (1970) in India. Except for some mutants selected by Doll, all the above-mentioned genotypes have reduced productivity and poor keeping quality. Both these shortcomings are being rectified and high-yielding high-lysine genotypes should soon become available for cultivation in different agro-climatic regions of the world. Safeguards against the selection of poor starch synthesisers in the artificially induced populations of high protein mutants have been suggested (Kaul, 1973).

It is common knowledge that, in cereals, the proteins of the germ are superior to those of endosperm. It has, therefore, been suggested that the biological value of such crops as maize and sorghum could be increased by selecting for large embryo size. In many of the existing Mexican races of maize the higher nutritive value can be directly ascribed to a favourable embryo/endosperm ratio. In sorghum this approach is already being followed (R. C. Picket, personal communication) and in the near future varieties with large embryos may become available.

It is now well established that considerable genetic variability in crop plants exists for the number of aleurone layers in the caryopsis and for the depth of the protein-rich aleurone zone. It is also known that in most of the cereals, proteins are distributed in a radial fashion getting more dilute towards the centre of the grain. These factors are of great significance in rice where milling removes the bulk of the high quality protein from the outer layers. Kaul *et al.* (1969) developed a microscopic screening technique for use in breeding rice varieties having deep-seated

Table 1.4. *Comparison of the effect of the Opaque-2 and Floury-2 genes in maize and the high-lysine gene in barley (whole seeds). After Munck (1972)*

	Normal* maize (g/16 g N)	Opaque-2* (normal maize = 100)	Floury-2* (normal maize = 100)	Normal† barley (g/16 g N)	High-lysine† barley (normal = 100)	High-lysine† barley (Floury-2 = 100)
Lys	3.0	167	160	3.4	123	86
His	2.6	135	112	2.1	103	75
NH_3	—	—	—	3.6	89	—
Arg	4.9	147	129	4.6	107	77
Asp	9.2	96	114	6.0	114	65
Tre	4.1	93	100	3.4	106	88
Ser	5.6	84	93	4.3	102	85
Glu	22.6	76	82	26.8	89	129
Pro	9.6	87	92	12.6	90	128
Gly	4.7	109	100	3.6	107	83
Ala	9.2	72	87	3.8	115	54
Cys	1.7	118	94	1.1	81	—
Val	5.7	91	100	4.8	110	92
Met	—	92‡	112‡	1.2	121	—
Ileu	4.2	81	95	3.7	105	98
Leu	14.6	64	82	6.7	106	59
Tyr	5.2	81	88	2.8	101	61
Phe	5.8	76	90	5.9	100	113
Try	0.7	185	—	—	—	—
Protein	9.0 %	117	189	15.7 %	108	101

* Nelson (1969).

† Mean of 35 normal and 27 high-lysine plants from a segregating F_2 population. Non-oxidised samples: Cys and Met absolute values not reliable.

‡ Approximation: normal maize = 2.3 g/16 g N.

proteins. Recently a similar pattern of radial distribution of proteins has been reported in oats (Youngs, 1972). Special milling processes will have to be perfected to obtain concentrated seed proteins (CSP) from low protein cereals (Burrows *et al.*, 1972). Rooney *et al.* (1972) could reconstitute flour, having up to 25 % protein, from a low protein dent variety of sorghum. These studies not only emphasise that 'polish' is of much higher biological value than the polished product but open up new possibilities of formulating high protein foods from low protein conventional crops. Furthermore, the necessity of breeding programmes aimed at deep-seated protein caryopsis is underlined. Regarding oats, it may be added here that this crop should regain its place in human nutrition.

A large number of less domesticated crop species are being evaluated for their direct, or indirect, use as sources of proteins. In some instances, e.g., teff (*Eragostis teff*), efforts are already underway to improve the components (Berhe, 1973). In some of the interior valleys of the Himalayan mountains buckwheat (*Fagopyrum sagittatum*) is the main source of protein and carbohydrates for human food. Considerable improvement in the protein properties of this crop are possible.

Triticale, the man-made cereal crop obtained by crossing wheat with rye (*Secale cereale*), has shown great promise with the protein content going up to 21 %. In a rat feeding experiment, Knipfel (1969) demonstrated that the PER (protein efficiency ratio) of triticale was superior to that of wheat or rye. He attributed this to a higher content of methionine and lysine in triticale. Poor grain filling and the presence of some growth inhibiting factors are the problems being attended to in this crop. Wild relatives of cereals are also being used to transfer single genes or small pieces of chromosomes into cultivated species in the hope of increasing their nutritive value.

Plant breeders have often improved plants long before the genetical and biochemical backgrounds of the improvement were fully understood. While much is yet to be understood about the physiological and biochemical control of grain protein synthesis and while nutritionists are still busy resolving controversies about the amino acid requirements of man, noteworthy progress has been made in breeding nutritionally superior genotypes of crop plants. When such improvements are not accompanied by reduced yields, they have the edge over other means of improving the supply and quality of food in that they do not involve recurrent expenditure on processing, transport, education and promotion. Finally, it needs to be mentioned that plant breeders are fully aware of their responsibility in minimizing the pre- and post-harvest losses of protein due to insects, moulds and faulty processing. These factors are taken into full consideration while breeding. The transfer of Opaque-2 gene into a dent background, breeding for hard textured wheats and diffusion of proteins into the inner cell-layers of endosperm in rice, are some examples in this category.

References

Anonymous (1973). *Indian Agricultural Research Institute Technical Bulletin*, (New series) 7. IARI, New Delhi.

Bansal, H. C. (1970). *Current Science*, **39**, 494.

Berhe, T. (1973). Proceedings of the Research Coordination Meeting, Neuherberg, on Nuclear Techniques for Seed Protein Improvement, pp. 297–303. International Atomic Energy Agency, Vienna.

Sources needing minimal processing

Burrows, V. D., Greene, A. H. M., Korol, M. A., Melnychyn, P., Pearson, G. G. & Sibbald, I. R. (1972). Food proteins from grains and oilseeds. Report of a Study Group appointed by the Hon. Otto E. Lang, Minister Responsible for the Canadian Wheat Board. Canadian Government.

Doll, H. (1972). Proceedings of the Study Group Meeting, Buenos Aires, on Induced Mutations and Plant Improvement, pp. 331–42. International Atomic Energy Agency, Vienna.

Eggum, B. O. (1973). Proceedings of the Research Coordination Meeting, Neuherberg, on Nuclear Techniques for Seed Protein Improvement, pp. 391–408. International Atomic Energy Agency, Vienna.

Hagberg, A & Karlsson, K. E. (1969). Proceedings of the Panel Meeting, Röstanga, on New Approaches to Breeding for Improved Plant Protein, pp. 17–21. International Atomic Energy Agency, Vienna.

Johnson, V. A., Mattern, P. J. & Schmidt, J. W. (1970). *Proceedings of the Nutrition Society*, **29**, 20–31.

Kaul, A. K. (1968). *Journal of the Post Graduate School, Indian Agricultural Research Institute*, **6**, 168–80.

Kaul, A. K. (1973). Proceedings of the Research Coordination Meeting, Neuherberg, on Nuclear Techniques for Seed Protein Improvement, pp. 1–107. International Atomic Energy Agency, Vienna.

Kaul, A. K., Dhar, R. D. & Swaminathan, M. S. (1969). *Current Science*, **38**, 529–30.

Kies, C., Fox, H. M. & Williams, E. R. (1967). *Journal of Nutrition*, **93**, 377–85.

Knipfel, J. E. (1969). *Cereal Chemistry*, **46**, 313–17.

Munck, L. (1972). *Hereditas*, **72**, 1–128.

Nelson, O. E. (1969). *Advances in Agronomy*, **21**, 171–94.

Rooney, L. W., Fryer, W. B. & Cater, C. M. (1972). *Cereal Chemistry*, **49**, 399–406.

Swaminathan, M. S., Austin, A., Kaul, A. K. & Naik, M. S. (1969). Proceedings of the Panel Meeting, Röstanga, on New Approaches to Breeding for Improved Plant Protein, pp. 71–86. International Atomic Energy Agency, Vienna.

Swaminathan, M. S., Naik, M. S., Kaul, A. K. & Austin, A. (1970). Proceedings of the Symposium, Vienna, on Improving Plant Protein by Nuclear Techniques, pp. 165–83. International Atomic Energy Agency, Vienna.

Youngs, V. L. (1972). *Cereal Chemistry*, **49**, 407–11.

For further information the reader is referred to various reports of the International Atomic Energy Agency (IAEA), Protein Advisory Group of the UN (PAG) and proceedings of American Chemical Society and American Association of Cereal Chemists.

2. Varietal improvement of seed legumes in India

L. M. JESWANI

In India, several 'seed legumes' commonly known as pulses have been traditional supplements to staple cereals. About 30% of the daily protein needs are contributed by pulses. Perhaps, no other country grows such a wide variety of pulses all the year round as India. There are about ten major pulse crops; Bengal gram or chick pea (*Cicer arietinum*) is the main one followed by red gram (*Cajanus cajan*), which is also known as arhar or tur. Other legumes include moong (or mung) bean or green gram (*Phaseolus aureus*), urd bean or black gram (*Phaseolus mungo*), moth bean (*Phaseolus acontifolius*), horse gram or kulthi (*Dolichos biflorus*), cowpea (*Vigna sinensis*), pea (*Pisum* spp.), lentil (*Lens esculentus*), khesari (*Lathyrus sativus*), and a number of other minor pulses. Efforts are being made to develop through genetic manipulation, short duration, high yielding and disease resistant varieties, which are not only rich in total proteins but are also rich in methionine and total sulphur.

Bengal gram

The Bengal gram or chick pea (*Cicer arietinum*), has two principal cultivated types; the brown or yellow-brown Deshi type and the white-seeded Kabuli type. Variation in their nutritive values are presented in Table 2.1.

Research work on varietal improvement on Bengal gram was initiated during 1930. The major emphasis was laid on single plant selection from bulk collections made all over the country. This work resulted in some of the very promising types, e.g., N.P.25, N.P.58, Chaffa, Dohad-Yellow, EB-28, etc. During the second stage from 1942 to 1966, major emphasis was laid on creating more variability by making further field collections in the country, introducing germplasm from other countries and following limited hybridisation programmes. Although types resistant to diseases were not then developed, a number of varieties were isolated which were better yielding and drought tolerant, e.g., R.S.10, G-24, T-3 and T-87. In the succeeding breeding programme major

Table 2.1. *Composition of two types of Bengal gram*

	Crude protein (%)	Ether extracts (%)	Crude fibre (%)	Ash (%)	Carbo-hydrates (% by difference)	Phos-phorus (mg/100 g)	Cal-cium (mg/100 g)	Iron (mg/100 g)
Kabuli type	21.64	5.78	5.49	2.67	64.42	305.8	167.4	8.36
Deshi type	20.91	4.56	10.06	2.69	61.78	308.8	231.1	6.90

emphasis has been laid on developing high yielding varieties resistant to diseases, particularly wilt and blight, and improvement in grain quality. Genetic sources of wilt and blight resistance have been identified and are being incorporated in some of the high yielding varieties; e.g., C-235, G-130, H-208, BGS-1, BGS-2, G-62-404 and Chaffa. Studies on the quality aspects have also revealed great variation in the protein and essential amino acid content as can be seen from Table 2.2. These materials are also being utilised in future breeding programmes.

Red gram

Red gram or pigeon-pea is the second most widely cultivated pulse in India. Based on morphological characters, two forms, namely *Cajanus cajan* var. *flavus*, commonly known as tur and *Cajanus cajan* var. *bicolor*, known as arhar, have been described. The former type includes the commonly cultivated varieties, which are relatively dwarf and bear yellow flowers and plain pods; the latter type includes most of the perennial types, which are generally late-maturing, tall and bushy varieties.

Research work on red gram was initiated as early as 1917. A number of varieties, e.g. RG-37, RG-72 (A.P.), S.A.1 (T.N.), Vijapur-49, T-15-15 (Gujarat), Thogari-2 (Mysore), Type-7, Type-17 (U.P.), B-7 (West Bengal), N-84 and N-148 (Maharashtra) and Gwalior-3 (M.P.) were developed either through single plant selections from local populations or through limited hybridisation programmes.

Wilt caused by *Fusarium udum* is a serious disease of red gram. Breeding for wilt resistance helped in isolating a few wilt resistant varities, e.g. NP (WR) 15, C-11, C-36, F-18 and F-52. These varieties showed 80–100 % resistance to wilt disease when grown in different agro-climatic regions of the country. The commercial varieties are of long duration (180–300 days) and are susceptible to frost which is

Table 2.2. *Protein, methionine and sulphur content of varieties of Bengal gram*

Variety	Protein content (%)	Methionine content (mg/g)	Sulphur content (mg/g)
S.1	20.10	2.40	3.10
S.6	19.25	3.30	2.50
S.8	20.80	3.10	2.90
S.12	20.56	3.00	3.55
S.13	18.60	3.10	2.90
BGS-1	21.35	3.30	2.45
BGS-2	19.52	3.00	2.00
P.1981	21.62	2.10	3.80
Commercial varieties			
C-235	20.40	2.60	1.75
T-3	19.30	2.00	1.98
N.P.58	20.50	1.96	1.98
N.59	18.60	1.90	1.92

Table 2.3. *Features of short duration varieties of red gram*

Variety	Maturity in days	Protein content (%)	Yield (kg/ha)	Yield (kg/day)	Protein (kg/ha)	Protein (kg/day)
S-3	165	21.12	1538	9.3	324.82	1.96
S-5 (Ageti)	150	20.75	1625	10.8	335.17	2.23
S-8 (Sharda)	140	20.68	1682	12.0	347.83	2.48
R-60 (Mukta)	170	21.30	1435	8.4	305.65	1.70
T-21	150	20.30	1260	8.4	255.78	1.50

common in north western parts of the country. Breeding work was initiated for developing short duration varieties which may escape the frost. These research efforts helped in developing early-maturing varities, e.g., T-21, S-3, S-5, S-8 and R-60, which mature in 120–170 days. Of these, S-3 and R-60 have exhibited resistance to stem rot and wilt diseases respectively and are being used as donor parents in future breeding programmes. Important features of some of the short duration varieties of red gram are presented in Table 2.3.

Breeding material on hand in red gram has also exhibited large variations in methionine and sulphur content, and this is also being exploited in the future breeding programme (see Table 2.4).

Table 2.4. *Methionine and sulphur content of varieties of red gram*

Variety	Methionine (mg/g)	Sulphur (mg/g)
P.2780	3.00	1.30
P.3758	2.40	2.50
P.4768	2.20	2.90
P.4415	2.60	1.90
P.4657	2.30	1.72
R.24	2.05	1.92
S.32	2.05	1.32
S.34	2.03	1.72
Commercial varieties		
T-21	1.33	1.52
C.11	1.80	1.70
N.84	1.55	1.50
T-15-15	1.60	1.50

Phaseolus group

Moong beans, urd beans and moth beans, are considered to be native to India, having been originated from *Phaseolus sublobatus* which grows wild in India.

Moong bean or green gram (Phaseolus aureus)

The research work on the improvement of moong beans was started in India in 1925 with large collections of seed samples from different districts of the country and also from Burma. Pure line selection from the local materials resulted in some promising varieties, e.g., GG-127, GG-188, Krishna-11, Khargone-1, Co. 1, Kopergaon, NP-23 and Jalgaon 781. Since the commercial varieties of moong bean are bushy, non-synchronous and late-maturing, the breeding programme was directed towards short duration types, like T-1, Krishna-11, Khargaon-1 (green seeded) and B-1 (golden seeded), which mature in 60–80 days. The succeeding breeding programme of hybridisation followed by reselection resulted in a set of promising varieties, e.g., T-2, T-44, T-51, Jawahar-45, R.S.4 and Gujarat-1. Selection programmes, seeking an ideal plant type with an erect upright habit and synchronous fruiting, resulted in the development of 'Pusa Baisakhi'. A further breeding programme has resulted in high yielding photo-insensitive, short duration varieties; e.g., S-8, S-9 and H-17-60. These have been found suitable both for summer and monsoon seasons and are being popularised in different multiple cropping programmes in the country.

Table 2.5. *Yield and protein content in high yielding varieties of urd bean*

Variety	Maturation time (days)	Protein content (%)	Yield (kg/ha)	Yield (kg/day)	Protein (kg/ha)	Protein (kg/day)
Pusa Sel. 1	80	27.0	1250	15.6	337.5	4.2
T-9	75	27.6	900	12.0	248.4	3.7
T.27	100	27.0	632	6.3	170.64	1.7

Moong bean is susceptible to a complex of virus and fungal diseases. Valuable genetic sources of resistance, e.g., L249, L355 and L661 for complex of virus diseases, and MG-32, MG-51 for cercospora leaf spot, have been isolated and are being used as donor parents in various breeding programmes.

Urd bean or black gram (Phaseolus mungo)

The earliest attempts to improve urd bean started in 1925, when 125 strains were isolated from the local bulks. Systematic improvement of urd was started in 1943. These efforts resulted in a number of promising varieties, both for dry areas, e.g., BG-379, B.R. 61, Mash-48, Mash 35-5, Khargone-3, T-27, T-65 and Sindh Kheda 1-1, and also for wetlands, e.g., ADT-1.

In the subsequent breeding programme, emphasis was laid on developing early-maturing varieties, resistant to yellow mosaic virus disease. Some of the germplasm lines, namely L-20, L-151 and L-190, were identified as the best sources of resistance to complex diseases. Pusa Selection-1 (Kamali), from a cross T-9 × L-151, and U.P.U. 1, by reselection from T-9, were developed; they are early-maturing and have a high degree of tolerance to diseases. Yield and protein content in some of the high yielding varieties of urd bean are presented in Table 2.5.

New breeding material also exhibits great variability with regard to methionine and sulphur content as can be seen from Table 2.6. This material is being exploited in future breeding programmes.

Moth bean (Phaseolus acontifolius)

A breeding programme on this crop was started in 1943 and 150 collections were made from the cultivated areas of the country. From single plant selections, two types, namely B-15 and B-18, were identified as good grain types and T-3 as a good fodder variety. Another variety,

Table 2.6. *Methionine and sulphur content of varieties of urd bean*

Variety	Methionine (mg/gm)	Sulphur (mg/gm)
New selections		
S.1	4.50	1.20
S.18	4.30	2.30
S.137	4.60	1.30
S.307-1-2	4.90	2.00
S.302-2-4	4.90	2.50
S.309-3-5	4.60	2.30
S.253-1-2	4.60	3.00
S.282-3-6	4.50	2.20
Commercial varieties		
Mash-35-5	2.50	1.73
Mash-48	3.00	1.73
T-9	2.65	2.65
B.R.61	3.50	2.20

No. 88, was identified as a better grain type, maturing in 120 days. These lines showed some improvement in yield, by 10–15 %, but no varieties resistant to diseases have been identified. Disease resistance and quality aspects are being considered in future breeding programmes.

Dolichos beans

Two major species of *Dolichos*, are commonly cultivated in India. One is *Dolichos lablab*, commonly known as walve or avare and the other *Dolichos biflorus*, known as horse gram or kulthi.

Walve or avare (Dolichos lablab)

Research work on improvement of avare has been carried out with the object of developing drought resistant, high yielding types with good quality pods. Some of the varieties, e.g., Co. 1, Co. 5 and Co. 6, have shown wide adaptability and are being popularised in rotation with late paddy in areas where winters are mild.

Horse gram or kulthi (Dolichos biflorus)

Very little work has been done on the improvement of horse gram; however, as a result of single plant selections from the local bulks, a number of varieties recording 15–20 % more yield than local bulks

Table 2.7. *Characteristics of some varieties of cowpea*

Variety	Origin	Maturation time (days)	Yield (kg/ha)	Protein content (%)	Methionine content (mg/g)
C-10	T-2 × Black eye	74	1550	25.1	1.6
C-13	T-2 × Black eye	65	1240	28.3	1.5
C-19	T-2 × Black eye	70	1650	26.5	1.4
C-20	T-2 × P-709	80	1750	27.5	1.6
C-152	T-2 × P-709	75	1810	26.8	1.3
FS-68	Haryana	80	1180	26.5	1.6
Commercial varieties					
K-11	M.P.	80	850	24.0	1.2
T-2	U.P.	120	948	25.2	1.0

have been developed. Some of the varieties, e.g., BGM 1-1, No. 35, D.B. 7 have been found promising. Variety BGM-1 exhibited a high degree of virus resistance.

Cowpea

A breeding programme for improvement of this crop (*Vigna sinensis*) has been in progress since 1940. A number of grain, fodder and vegetable varieties have been identified from time to time largely from collections made within the country or from abroad. Of the grain types, N.P. 2, N.P. 7, C-32, T-1, K-11, K-14; of the fodder varieties, Russian giant, T-2, K-782, K-397, C-55, C-322; and of vegetable types, Pusa Phalguni, Pusa Barsati, Pusa Dofasli and FS-68 have been identified.

Realising that all the grain types are shybearers and late-maturing, the breeding programme was reoriented to develop short duration, high-yielding and disease resistant varieties. A number of varieties, e.g., C-10, C-13, C-19, C-20 and C-152 combined short duration and other agronomic characters (see Table 2.7). Some of the strains (C-13, C-20 and C-152), exhibit a high degree of tolerance to virus and bacterial blight. A few strains, e.g., No. 585, No. 700 and No. 782 have shown resistance to *Microphomina phaseolia*.

Pea or matar

There are two main types of cultivated pea (*Pisum sativum*), namely the large, smooth or wrinkled-seeded garden pea and the small, round or dimpled-seeded field peas. While the former type is used as a table variety, the latter is used as pulse, whole or split.

Garden pea

A breeding programme on the garden pea was initiated at the Indian Agricultural Research Institute in the thirties. Through single plant selection, the medium-tall, wrinkled-seeded variety NP-29 was developed which is still popular in the country for its quality. During the same period, green-seeded Hara Bauna and white round-seeded Lucknow Poniya were popularised for general cultivation in northern India. In central India, where the winters are comparatively short, the variety Khapar Kheda became more popular. In the warm-temperate zone around the Himalayas, a smooth, white-seeded variety was popularised under the name Kala Nagini or Kanawari. In recent years, a few more varieties, e.g., Early Badger, Boneville and Perfection with very attractive pod size have been introduced for general cultivation.

Field peas

Several types of field pea are grown in different agro-climatic regions of the country. A high yielding bold-seeded white variety, Type 163, was developed in 1950. In the succeeding breeding programme, a few varieties, e.g., T. 6113 and T. 6115, were developed by hybridisation. A few more white-seeded types BR-2 and BR-118 were released for the plains in Bihar state and BR-178 for the hilly areas.

Leaf miner and powdery mildew cause heavy losses. Genetic sources of resistance have been identified and breeding programmes have been reoriented to develop high yielding resistant varieties of pea.

Lentil

Varietal improvement programmes for this species (*Lens esculentus*), were initiated in India in 1924 by collecting mixed samples bought in bazaars all over the country. Single plant selections were picked up from the bulk population and sixty-six types were isolated. Some of these varieties, e.g., N.P. 11, N.P. 47 (IARI), T-36, T-8 (U.P.), L-9-12 (Punjab) and B.R. 25 (Bihar) are promising. Paddy–lentil is a common rotation in India. A breeding programme was initiated to develop early maturing, high yielding varieties suitable for late sowings under soil moisture stresses. A number of such quick maturing varieties, suitable for late sowings were developed as listed in Table 2.8.

Lentil (also known as masoor or masari), is susceptible to rust and wilt diseases. Genetic sources of resistance have been isolated which are

Table 2.8. *Yields of quick maturing varieties of lentil*

Variety	No. of days in the field	Calculated yield (kg/ha)	Yield (kg/day)
Pusa-1	70	1840.00	26.3
Pusa-4	74	1717.80	23.2
Pusa-6	72	1892.00	26.3
Commercial varieties			
L-9-12 (Punjab)	110	1825.00	16.5
T-36 (U.P.)	94	1743.00	18.5

Table 2.9. *Yield, protein and neurotoxin contents of khesari varieties*

Variety	Yield (kg/ha)	Protein content (%)	Neurotoxin content (%)
New strain			
Pusa 10-1	683.2	22.56	0.18
Pusa 24	664.5	21.65	0.19
Pusa 396	748.3	21.12	0.14
Pusa 648	569.5	26.19	0.22
Pusa 719	475.8	23.25	0.20
Commercial varieties			
T-12	403.6	22.32	1.28
L.C.76	658.4	23.00	1.65
Rewa-1	728-8	23.00	0.98

being used in breeding programmes for developing short duration, high yielding varieties resistant to diseases.

Khesari or teora

The consumption of this pulse (*Lathyrus sativus*), in large quantities leads to lathyrism because of the presence of β-N-oxalyl amino alanine (βOAA). It is a very hardy crop and comes up well even under water-logging and extreme drought conditions. Therefore in areas which are completely dependent on the monsoon, farmers insist on growing it.

Two forms of khesari are common; one with smaller seed size known as lakhori and the other with bigger seed size known as lakh. In the earlier years (1940–60), varietal improvement of khesari was undertaken primarily to increase yield by making single plant selections from the local bulks of lakhori/lakh. Some of the varieties, e.g., T2–12, L.C. 76, Rewa-1, Rewa-2, were released for general cultivation. The major problem of lathyrism, however, remained unsolved. Efforts by the Government to discourage its cultivation also met with no success.

During 1966–7, the Indian Agricultural Research Institute initiated a breeding programme to identify sources with low βOAA content and develop high yielding varieties free from βOAA. As a result, a few strains having a low level of the neurotoxin compound (βOAA) have been isolated. Table 2.9 presents the neurotoxin levels and yield of some *Lathyrus* lines which appear quite promising and safe.

For further information on seed legumes the reader is referred to:

New vistas in pulse production. Indian Agricultural Research Institute Bulletin No. 4, New Delhi, 1971.

Nutritional improvement of food legumes by breeding. Protein Advisory Group of the UN, New York, 1973.

3. Minor food seeds

D. A. V. DENDY, BERNICE EMMETT & O. L. OKE

Eighteen of the many hundreds of minor food seeds have been chosen for individual discussion. Most are of only local interest but all, notably quinoa, could become important sources of protein throughout the world. Though many traditional crops have been or are being supplanted by the popular cereals and legumes, there will remain regions where the minor seeds will continue to play an important role in nutrition. When there has been an amount of research on these minor seeds comparable to that on cereal and legume seeds, the yields will probably increase. This is especially so for those seeds already adapted to extreme climatic conditions such as high altitude and/or low rainfall, and for seeds such as quinoa with better nutritional properties than the well-known cereals.

Bambarra groundnut

Grown in West Africa (Masefield *et al.*, 1969), the Bambarra groundnut (*Voandzeia subterranea*), prefers light soils, with a warm to hot climate, with rain during planting but dry during harvest. The crop takes 100–180 days to mature. The seeds are sown with a spacing of 30–45 cm between seeds and 15–54 cm between rows, at a seed rate of 35–60 kg/ha. The yield is 300–800 kg/ha (FAO, 1961). The Bambarra groundnut has the following chemical properties: moisture 10 %, protein 21.1 %, carbohydrate 53.5 %, fat 6.5 %. It has a high lysine content (6.4 g/16 g N) (O. L. Oke, personal communication), and fairly high levels of other amino acids.

Buckwheat

Buckwheat (*Fagopyrum* spp.) a member of the Polygonaceae family, is grown in the USA, USSR, Canada and Africa, but grows best in areas which have a fairly cool, moist climate. The principal species of economic importance are *F. esculentum*, *F. tataricum* and *F. emarginatum*. There are many varieties of *F. esculentum*. They grow on infertile soil but prefer sandy, well-drained soil and have a high tolerance to soil acidity. Buckwheat is very sensitive to cold and will be killed by a slight frost. In the cool areas it takes about 70–84 days to mature. The seed is either broadcast or drilled 2.5–5 cm apart in rows 17.5–38 cm apart, at a seed rate of 25–40 kg/ha. The average yield is 1000 kg/ha but under very favourable conditions it may yield 3800–4000 kg/ha (FAO, 1961). Buckwheat has a moisture content of 13.0 % and a protein content of 12.0 %, comparable to that of some wheats. There is also present a trace of vitamin A, 0.3 mg of vitamin B_1, and 0.3 mg of vitamin B_2 per 100 g edible portion. Results available for mineral content are: calcium 30 mg/100 g sample and iron 3.0 mg/100 g sample. Its carbohydrate content is about 71.0 % and the fat level is about 2.0 % (Platt, 1962). The percentage of total nitrogen present as amino acids is 79.0. Buckwheat is not deficient in any essential amino acids (Van Ethen *et al.*, 1963).

Carob or locustbean

This plant, (*Ceratonia siliqua*), is grown in the sub-tropics, in southern USA, Cyprus, Australia and South Africa. Three to four years after planting the grafted trees, yields up to 480 kg/tree, or 8000 kg/ha are obtained. In Cyprus it is grown for the production of gum from the endosperm. The gum, which contains up to 60 % protein, is ground and used in feedstuff. The bean contains 21 % of protein and a fairly low fat level of 1.5 % (Douglas, 1972*a*, *b*); 92 % of the total nitrogen is protein nitrogen (Van Ethen *et al.*, 1963).

Chenopods, in particular quinoa and canihua

Though the chenopods occur worldwide and are gathered for food throughout the Americas, it is only in the Altiplano region of the Andes that they are cultivated as a major item of diet and, indeed, have been so cultivated since before the Inca Empire. Two species, quinoa (*Chenopodium quinoa*) and canihua (*C. pallidicaule*), are of major importance. These are herbaceous annuals growing to a height of 1.2–1.8 m cultivated chiefly in Peru, Bolivia and Argentina. They thrive best at altitudes of 1500–3600 m. *C. pallidicaule* can be grown at higher altitudes than *C. quinoa* because its vegetative period is 135–145 days compared with 165–172 days for quinoa. Both species tolerate light frosts and give good seed yield with only 300 mm of rainfall, but young plants are susceptible to severe drought and frost. They are adapted to a wide range of soils but prefer those rich in lime, potassium and magnesium. They tolerate soluble salts in the soil and can grow where other cereals do poorly (FAO, 1961).

It has been estimated that 5000 ha or about one-sixth of the cultivated area of the Altiplano is planted with quinoa. The yield is variable, 350 to 800 kg/ha, but reaches 4000 kg/ha under experimental conditions (FAO, 1950; White *et al.*, 1955). It has been estimated that yearly production in Peru alone is 4000 tonnes of quinoa and 10000 tonnes of canihua (White *et al.*, 1955). The seed is threshed from the harvested plant and because it is very small (1–2 mm in diameter) and of irregular shape, little effort has been made to mechanise harvesting. The crop is grown mainly by peasants and is a major item of their diet, but its importance in the diet of townspeople has declined since the 1840s when the Bolivian President provided subsidies to quinoa growers to encourage it as a cash crop. A gruel is apparently still used by townspeople in times of sickness (D. F. Ridgeway, personal communication).

The seeds contain a poisonous but water-soluble saponin which must be leached out with water or destroyed by toasting before the seeds are cooked and eaten. Quinoa contains around 12 % protein (N × 6.25) 5 % fat, 60 % starch, and canihua 14 %, 4 % and 52 % respectively (Wolfe *et al.*, 1950; White *et al.*, 1955; Quiros-Perez & Elvehjem, 1957; de Bruin, 1964). The protein is of remarkably high quality, containing 7 g/16 g N lysine (6 in canihua) (White *et al.*, 1955). Unlike the cereal grains, chenopod starch is stored in the perisperm and is similar in physical properties to cornstarch (Wolfe *et al.*, 1950). The vitamin content compares very favourably with true cereals. For example, the niacine content is 9.6 mg/100 g (Mazzocco, 1934; FAO, 1950).

Nutritional tests indicate that chenopods are valuable sources of protein if correctly prepared by washing out the saponins. Rat growth studies indicate that quinoa and canihua are both superior to wheat and equal to, if not better than, dried whole milk (White *et al.*, 1955).

In general, the difficulty of removing the saponin has limited the use of quinoa for diluting wheat flour. Apparently reasonably good bread can be made from strong wheat with up to 20 % quinoa, though further work on this would obviously be desirable. Chenopods have also been used for poultry feed.

Other chenopods to attract interest recently are the Prickly saltwort or Russian thistle (*Salsola kali*), Burning bush or Summer cypress (*Kochia scoparia*) and Garden orache (*Atriplex hortensis*). These seeds have been considered for poultry feed and nutritional studies have been carried out, but none has been cultivated for human food (Larmour & MacEwen, 1938; Coxworth *et al.*, 1969).

In view of the good nutritional value of chenopods and the fact that they will grow in semi-arid regions at high altitudes, it is clear that the growing of these species and the breeding of high-yielding varieties should be encouraged.

Cowpea

The cowpea (*Vigna* spp.) is grown in Africa, southern USA and China (Masefield *et al.*, 1969). It is frost sensitive and will grow in a wide range of non-alkaline soils. The seeds are sown with a spacing of 45 × 45 cm, at a seed rate of 30–60 kg/ha. The seed takes 3 months to mature. The yield is 500–800 kg/ha, though, some improved varieties produce 1000–1500 kg/ha (FAO, 1961).

The cowpea has a high protein and carbohydrate content, 22.0 and 66.0 % respectively, a fat content of 1.5 %, 180 mg/100 g calcium, and 22 mg/100 g iron (Platt, 1962). Total N as amino acid is 80 %; it has a high lysine content of 6.4 g/16 g N, and a fairly high phenylalanine content of 6.0 g/16 g N (Van Ethen *et al.*, 1963).

The cucurbits

The cucurbits are widely grown in warm, temperate regions and in the drier parts of the tropics. They like well-drained, fertile soil, with even rainfall. Maturity occurs in about 90–120 days. They are frost sensitive but will survive in a cool, moist environment; most species are sensitive to extreme heat. To prevent build-up of soil-borne diseases cucurbits should not be grown every year without alternating another crop. Seed is usually sown at about 3–5 kg/ha (FAO, 1961). Three species are used partly for their edible seeds, but many more species can yield edible seeds in addition to the flesh, which is normally eaten. Yield of seed is in the range 300 to 3000 kg/ha. Seed contains around 35 % protein (dry wt) and up to 50 % oil, rich in unsaturated fatty acids (Jacks *et al.*, 1972).

Pumpkin (Cucurbita pepo)

In West Africa the seeds are removed by crushing the fruit and leaving it in water (FAO, 1961). The seeds, whose protein content is about 27.9 %, are boiled and eaten. The seed yield is 600–800 kg/ha.

Melon (Citrullus vulgaris)

Grown in India, Africa and parts of Asia, especially in the drier regions, the seed has a protein content of about 32%. The seed yield is 110–

220 kg/ha. It is used in a ground form as a condiment in soups, or mixed and cooked with vegetables.

Fluted pumpkin seed (Telfairia occidentalis)

Grown in West Africa, this species has a protein content of about 28 %. It is cooked and eaten like breadfruit or walnuts and can be ground up and added to soup.

Hausa groundnut

Kerstingiella geocarpa is a member of the legume family grown in the drier parts of West Africa in light loam or sandy soils (Masefield *et al.*, 1969). No figures are available for yield, but they are said to be low (Irvine, 1969). Maturity is reached in 4–5 months. The kernel, which has a protein content of about 20.5 %, may be boiled, salted and eaten with shea-butter. The Hausa groundnut has a high fat content (58.0 %), and a fairly high lysine content (6.6 g/16 g N), but is low in tryptophan (0.8 g/16 g N) (O. L. Oke, personal communication).

Lupin

This legume (*Lupinus* spp.) grows in Europe, North Africa, western Australia, Peru, Bolivia and the USA. It prefers cool weather with about 900 mm annual rainfall, and a medium-rich light soil. It can grow at high altitudes; *L. luteus* will grow at up to 3300 m. The lupin matures in 100–160 days. The seeds are sown at a spacing of 60–90 cm, at a seed rate of 60–160 kg/ha, depending on species (FAO, 1961). It has a protein content of 35.0–40.0 % and a fat content of 6.0–12.0 %, both quantities being species dependent.

According to van Ethen *et al.* (1963), *L. luteus* has a high tyrosine and arginine content, 7.0 g/16 g N and 13.9 g/16 g N respectively, but a low phenylalanine (3.6 g/16 g N), methionine (0.5 g/16 g N) and threonine (2.9 g/16 g N) content. The seed yield given by FAO in 1961 was 800–1000 kg/ha but recent Spanish yields are larger. Lupin seed is ground up to a meal which, at present, is used only as an animal foodstuff.

Nigerseed

Guizotia abyssinica, commonly called nigerseed or nug (Chavan, 1961) is grown extensively in India and Ethiopia, and to a limited extent in

East Africa and the West Indies. It requires a moderate rainfall, not above 1000 mm, and will grow at altitudes of up to 2000 m in a wide range of soils. The plant requires moist conditions during the first two months after sowing and fruits after $3\frac{1}{4}$–5 months. Nigerseed is either broadcast at a rate of 10 kg/ha, or 5 kg/ha is sown in rows 46 cm apart. The seed yield varies from country to country: 300–400 kg/ha in India and Ethiopia, and 600 kg/ha in Kenya (FAO, 1961). It contains 6.2 % moisture, 17 % protein, 17 % fat and 37 % carbohydrate (Platt, 1962). The PER (protein efficiency ratio) is 1.5 (O. L. Oke, personal communication) and it is slightly deficient in lysine (4.4 g/16 g N) and methionine (1.5 g/16 g N). The mineral content depends on the area in which it is grown.

Teosinte

Teosinte (*Euchlaena mexicana*) can be grown in areas with rainfall between 600 and 1800 mm. The seeds are sown with a spacing of 60 cm between rows at a seed rate of 25–30 kg/ha and yields are up to 1200 kg/ha (FAO, 1961). It has a moisture content of 11.0 % and a protein content of 14.0 %; fat is 3.0 % and carbohydrate 62 %. Also present are traces of vitamins (Platt, 1962). Total nitrogen as amino acids is 72.0 %. Teosinte is deficient in lysine and arginine (1.7 g/16 g N and 3.3 g/16 g N respectively), but it has a high leucine level of 13.7 g/16 g N (Van Ethen *et al.*, 1963).

West African locust bean

This bean (*Parkia filicoidea*), is grown in North Nigeria and south east Asia. It has a moisture content of 10 %, 30 % protein and 20 % fat. The locust bean is rich in lysine (7.4 g/16 g N) and in leucine (8 g/16 g N), but poor in methionine (1 g/16 g N). After fermenting, drying and mixing with wood ash, the final product is eaten as a relish with various dishes (O. L. Oke, personal communication).

Winged beans

Winged beans (*Psophocarpus tetragonolobus*) are grown in west Africa and tropical Asia. Sown with a seed spacing of 2 × 260 cm, the 'bean' matures 45 days after fertilisation. The yield is about 290 kg/ha. The winged bean is a useful source of protein (37.3 %), and has a fat content of 18.0 %. Unripe seeds are used in soup; ripe seeds may be roasted (Pospisil *et al.*, 1971).

Yam bean

Ethiopian in origin, (*Sphenostylis stenocarpa*) is common to many parts of tropical and equatorial Africa (Irvine, 1969). It has a protein content of about 19.2 %, fat 1.1 % and carbohydrate about 67.4 %. It also contains 55 mg calcium per 100 g. The yam bean is not deficient in any essential amino acids while the lysine content is high at 6.8 g/16 g N (O. L. Oke, personal communication).

References

Chavan, V. M. (1961). *Niger and safflower*. Examiner Press, Bombay.

Coxworth, E. C. M., Ball, J. M. & Ashford, R. (1969). Preliminary evaluation of Russian thistle, kochia and garden atriplex. *Canadian Journal of Plant Science*, **49**, 427–34.

de Bruin, A. (1964). Investigation of the food value of quinoa and canihua, *Journal of Food Science*, **29**, 872–6.

Douglas, S. J. (1972*a*). Tree crops for food forage and cash. Part 1. *World Crops*, **24**, 15–19.

Douglas, S. J. (1972*b*). Tree crops for food, forage and cash. Part 2. *World Crops*, **24**, 86–9, 97.

FAO, (1950). *Agriculture in the Altiplano of Bolivia*. FAO Development Paper No. 4. FAO, Washington.

FAO, (1961). *Agricultural and horticultural seeds*. FAO Agricultural Studies No. 55. FAO, Rome.

FAO/USDHW. (1968). Food composition tables for use in Africa. FAO, Washington.

Irvine, F. R. (1969). *West African crops*, vol. 2. Oxford University Press, London.

Jacks, T. J., Hensarling, T. P. & Yatsu, L. Y. (1972). *Economic botany*, **26**, 135–41.

Larmour, R. K. & MacEwan, J. W. G. (1938). The chemical composition of Russian thistle. *Journal of Scientific Agriculture*, **18**, 695–9.

Masefield, G. B., Harrison, S. G. & Wallis, M. (1969). *Oxford book of food plants*. Oxford University Press, London.

Mazzocco, P. Vitaminas de la quinoa. (1934). *Sociedad Argentina de Biologia*, **10**, 367.

Platt, B. S. (1962). *Tables of representative values of foods used in tropical countries*. HMSO, London.

Pospisil, F., Karikari, S. K. & Boamah-Mensah, E. Investigations of winged beans in Ghana. *World Crops*, **23**, 5.

Quiros-Perez, F. & Elvehjem, C. A. (1957). Nutritive value of quinoa proteins. *Journal of Agricultural and Food Chemistry*, **5**, 538.

Van Ethen, C. H., Miller, R. W. & Wolff, I. A. (1963). Amino acid composition of seeds from 200 angiospermous plant species. *Journal of Agricultural and Food Chemistry*, **11**, 399–410.

Sources needing minimal processing

White, P. L., Alvistur, E., Dias, C., Viñas, E., White, H. S. & Collazos, C. (1955). Nutrient content and protein quality of quinoa and canihua. *Journal of Agricultural and Food Chemistry*, **3**, 531.

Wolfe, M. J., MacMasters, M. M. & Rist, C. E. (1960). Some characteristics of the starches of three South American seeds. *Cereal Chemistry*, **27**, 219.

4. Vegetables

F. W. SHEPHERD

The term vegetable is used here to describe plants and parts of plants, other than ripe fruits and seeds, that are eaten after cooking; salads are similar parts eaten raw. Even with this limitation, the number of species that might be discussed is enormous. But the number that it is profitable to discuss is very small because little fully quantitative work has been published. Publications from which it is possible to calculate the amount of edible protein produced per hectare in a day or in a year are exceptional. This is surprising because such figures are readily available for most other crops, because vegetables are the source of about the same amount of food protein as fish, and because they are extremely productive. Annual protein yields in some experiments have reached 7 tonnes/hectare: this is so much more than can be produced by any other method of farming that vegetable production deserves greatly increased publicity and detailed investigation.

Vegetables are discussed here as sources of protein. They are more often discussed as sources of vitamins and minerals and they are probably most important in that role in infant feeding. Because of the limited capacity of the infant stomach it would be difficult for a child to eat enough leafy material to contribute more than 1 or 2 g of protein, but the amount often eaten could supply the whole requirement of carotene (pro-Vitamin A). Unfortunately, the modern 'industrial' trend may entail prolonged transport of the crop and pretreatments such as heating, drying or freezing, during which vitamins are, to varying extents, destroyed.

In general the number of species currently consumed is less than it was in the past. In most countries the stage when wild plants were collected for use as important contributions to the diet or as additional flavourings for uninteresting or monotonous diets has long since passed, although some small contributions are still made by wild plants in some rural communities. The further stage of home gardening or small scale production for local consumption is also rapidly passing. This is due in part to the increasing demand for processed or semi-processed vegetables, and other foods, in countries where both husband and wife are working outside the home and where meals are more often prepared in bulk for communal feeding. Relatively few vegetables have been developed to permit their transport and sale in these partially prepared forms; production of these vegetables is increasing whereas production of many others is decreasing. Dried, canned and frozen peas and beans are examples of vegetables which have been developed in this way. Others, such as some of the brassicas and spinach substitutes are not yet so developed, are costing more to produce and market, and are becoming less generally available. At the same time a decreasing labour force in the developed countries has led to increased mechanisation of production and harvesting and it is only those crops which can be grown by these methods that are likely to survive as economically viable crops, while those less easily mechanised become more costly and, therefore, find a smaller demand.

At present the effect of these losses on the diets in developed countries is probably unimportant but in less-developed countries such losses may become serious. For though, in total, the various vegetables and salads contribute little to the overall protein consumption of the world, their importance in some regions is vital. Where other protein is plentiful their main value is in the minerals and vitamins which they contain; elsewhere they are the basis of the whole diet. In fact only some 4 % of the protein consumed in the world comes from these vegetables and salads. Within this overall total the amount of protein supplied by vegetables varies very considerably from one country to another. Table 4.1 (FAO, 1971) shows the amounts consumed in a few countries only, but it indicates the wide range between those mentioned. Nor do other proteins necessarily complement vegetable protein: in some countries such as India, where the consumption of vegetables is very low, supplies of animal protein are equally low. On the other hand in countries such as the UK and the USA where vegetables are consumed in some quantity, the consumption of animal protein is also adequate.

The total per capita consumption of vegetables is slightly higher in most countries as home production must be added to these figures, but they give an indication of the situation in a range of countries.

It is to be regretted that in many countries the range of species and the total amount being grown and consumed as vegetables and salads is decreasing and that the amount of research on the production of all but a few is limited or non-existent. The economic and social developments which have increased production of a few vegetables and reduced consumption of others in the developed countries is unlikely suddenly to change, but the overall effect on the diet and health of these nations is unlikely to be serious in the short term. There are, however, some

Table 4.1. *Grams of protein, per head per day, supplied by commercially grown vegetables*

Portugal	7.8	Ceylon	2.0
Italy	5.1	Denmark	1.7
Japan	5.1	Nigeria	1.1
France	5.0	Guatemala	1.0
USA	3.8	Brazil	0.5
Israel	3.7	Mexico	0.4
Chile	3.3	Venezuela	0.2
UK	2.6	India	0.1

indications of dissatisfaction with the existing supplies of the small variety of processed or prepared vegetables and that this is leading to a demand for less common vegetables which, in Europe, is met by imports from warmer climates, while in North America there is also some reversion to the production of non-commercial kinds in home gardens.

It is in the less-developed regions that the effect of reduced or insufficient vegetable supplies can be most serious and it is here that there is often room for improvement in supply and quality. Governments and outside agencies are usually aware of the situation and valuable work is in progress on the potentialities of new crops, increasing the production of the old ones and publicising the value of crops found to be successful in local trials. Some examples of yields and values are given in Table 4.2 but it must be emphasised that in this field of study more work is needed on all aspects of production and preparation of local crops and potential new ones.

Not only is it necessary to study all the factors affecting vegetable production and to search for the most suitable vegetables for each situation, but there is also an urgent need to study the established and possible new crops, and to measure both the yields of protein, and other nutrients, from each hectare and the time taken to produce these yields. Most research results quote yields in terms of weight of harvested or marketable crop; far too few give the weight of protein, or even of nitrogen or consumable material, and the time from seed sowing to maturity is often not given.

The cultivator needs information about the most suitable species and cultivars and their consumable or marketable yields. He also wants guidance on their behaviour in the soil and climate of his locality, with information on cultivation methods and the control of diseases, pests

Table 4.2

Latin name	Common name	Duration (wks)	Part eaten	Wet weight as harvested (t/ha)	% dry matter in edible part	% 'protein' in dry matter	Rate of 'protein' synthesis (kg/ha · day)
Allium porrum	Leek	12	Leaf bases	40	15	30	21
Allium cepa	Onion	25	Bulb	50	13	20	8.5
Beta vulgaris	Beetroot	13	Root	35	13	14	7
Brassica oleracea	Cauliflower	30	Immature flower	30	10	30	4.3
Brassica oleracea	Savoy	30	Packed leaves	25	7	20	1.7
Brassica oleracea	Cabbage	30	Packed leaves	50	7	20	3.4
Brassica oleracea	Brussels sprouts	30	Axillary buds	25	7	20	1.7
Lactuca sativa	Lettuce	12	Leaves	16	6	23	2.6
Solanum lycopersicum	Tomato	25	Fruit	250	6	16	11

All these figures are taken somewhat arbitrarily from the extensive normal range in UK. The column labelled 'Duration' is the time in weeks for which the crop occupies a field; with some crops e.g., Brussels sprouts, much of this is winter when there is little growth. The yield is what an expert market gardener can reasonably expect. The proportion of this yield that is usually considered edible varies greatly: it is close to 100 % with beetroot and tomato and about 50 % with cabbage. The value for 'protein' is based on the nitrogen content; the proportion that is true protein varies not only between species but, within a species, according to maturity. The non-protein nitrogen is not, however, nutritionally negligible. Taking these factors into consideration, the actual rates at which edible true protein is being synthesised will be between half and two-thirds the rate given in the last column.

and weeds. The consumer, administrator and economist, however, also wish to know more of the food value of the edible portion of the crops and the time taken to produce it. With this information diets and production can be planned in the best interests of all.

Some of this information is available in scattered and unrelated forms but much remains to be obtained, and it would be helpful if all these factors were studied and adequately reported in future work on vegetable production.

Table 4.2 lists some of the vegetables which are grown in the mainly temperate climates of central and northern Europe, much of North America and in Australasia. In these countries climate rather than soil or pests, limits growth. The lack of light and adequate temperature during the winter months prevent all-the-year round production of most crops and thus makes the overall total production less than can be achieved in warmer climates that have more evenly distributed sunlight. If the production of the greatest quantity of protein or other nutrients were the only objective, the total yields could be far higher than is actually achieved. Consumer demand and, therefore, the economics of production often reduce total and saleable yields where demand is for the qualities of moderate size, shape, appearance and flavour instead of total quantity. For example, the maximum yield of carrots, which contain about 1 g of protein in each 100 g of fresh consumable product, can be achieved by growing the potentially largest sized cultivars at the optimum spacing and leaving them to grow up to 16 weeks from seed sowing to harvest; then some 50 tonnes of marketable product may be obtained from each hectare. There is, however, a considerable demand for much smaller roots which are sold either fresh or canned. These small roots are produced by selecting cultivars which mature at a smaller size, by sowing the seed more closely than would produce the maximum yield per unit area and by lifting the crop at an earlier stage in its progress towards maturity. In extreme cases yields early in the season may then be as low as 5 tonnes per hectare and although earlier harvesting may then allow a second crop to be grown, the total production of food is considerably lower than if maximum production were the aim. Similarly current demand is increasingly for small-sized Brussels sprouts, either frozen or in cans. This demand has induced plant breeders to produce smaller sprouts all at about the same time in order to allow mechanical harvesting of the crop when it is at the right stage of maturity. Plants of these newer kinds may be planted closer together than the older ones but their total yields by the newer methods may be no more than 5 tonnes to the hectare compared with 25 tonnes obtained from older methods and cultivars.

The proportion of the nitrogen in a vegetable that is protein nitrogen is seldom stated. Judging from the results that have been published, 70 to 80 % of the nitrogen is usually protein nitrogen. There is at present even less information about protein yields from tropical than from temperate-zone vegetables, but the establishment of international vegetable research institutes in Taiwan and elsewhere should soon improve matters. In the meantime a few scattered observations may be noted. *Basella alba* in Sri Lanka synthesises about 20 kg of true protein/ha·day (Gunetileke in Pirie, 1971). That is also the rate for *Amaranthus cruentus* in Nigeria (Schmidt, 1971), and it is a rate that is probably reached by *Ipomoea aquatica* in Hong Kong (Edie & Ho, 1969). The rates for *Corchorus olitorus* and *Hibiscus esculentus* are about 10 kg of protein/ha·day.

References

Edie, H. H. & Ho, B. W. C. (1969). *Ipomoea aquatica* as a vegetable crop in Hong Kong. *Economic Botany*, **23**, 32.

FAO (1971). *Production yearbook* 25. FAO, Rome.

Pirie, N. W. (1971). *Leaf protein: its agronomy, preparation, quality and use*. IBP Handbook 20. Blackwell Scientific, Oxford.

Schmidt, D. R. (1971). Comparative yields and composition of eight tropical leafy vegetables grown at two soil fertility levels. *Agronomy Journal*, **63**, 546.

5. The *Spirulina* algae

N. W. PIRIE

Spirulina, one of the Cyanophyta, is a helicoidal, blue-green alga, 0.2 mm long. It grows in highly alkaline water – this gives it some important advantages.

(1) Natural alkaline lakes are usually found in the arid zones; the production of an alga rich in protein would obviously be useful in these zones because alkaline waters are not suitable for irrigation.

(2) The high pH of the culture medium ensures that abundant carbon dioxide for photosynthesis is present at all times. By contrast, *Chlorella* grows in an acid medium and must be supplied with carbon dioxide continuously.

(3) The alkalinity prevents the invasion and development of other micro-organisms, which do not tolerate a high pH. Thus, in a semi-natural condition, or in culture ponds, *Spirulina* lives in an environment that contains few potentially harmful organisms.

The special merit of *Spirulina* is that it contains from 64 to 70 % of high quality protein on a dry weight basis. This percentage is greater than that of any other product in its natural, unconcentrated, state.

These advantages explain why *Spirulina* is a traditional food among people living near saline alkaline lakes where the alga grows spontaneously. In Lake Tchad, the Kanembou consume algae as dry cakes called 'dihe', which they eat two or three times a week with millet, seasoning it with spiced sauce. In Mexico, the Aztecs of Texcoco, cradle of the Nahuatl civilisation, knew it by the name of 'tecuitlatl' and used it as a complement to maize.

The annual yield of *Spirulina* (dry matter) cultivated in a synthetic medium in a specially designed pond can reach 40 tonnes/ha when the conditions of culture are favourable. In the semi-natural conditions of the Sosa Texcoco solar evaporator (pH 9.8), the yield is 10 tonnes/ha.

Concentration and filtration of the *Spirulina*

Because of their helical shape, the individual *Spirulina* cells, from artificial or natural culture ponds, form entangled clumps that can be concentrated by letting the suspension flow over an inclined filtration surface before vacuum filtration. Filtration can be carried out on a rotary drum or horizontal filter similar to the filters used in the paper industry. The *Spirulina* is washed on the same filter in order to eliminate the salts retained in the cake: the lake water at Texcoco contains 30 g of solids per litre. The filter cake can be dried on a rotary drum or a spray dryer.

Both methods give excellent results and the dried product can be stored without difficulty.

Nutritional qualities of the *Spirulina*

There are small morphological differences between *Spirulina* from Tchad and from Mexico, but no differences have been found in the nutritional qualities of the *Spirulina* grown in the traditional way in Tchad, in a semi-natural way in Mexico, or in an artificial medium in the laboratory. Dried alga contains 65 to 70 % protein, 6 to 7 % lipid, and 4 % nucleic acid. The amino acid composition is similar to that of FAO 'reference protein' except for a slight cystine deficiency.

Tests on rats show BV = 72; NPU = 61–63; PER = 2.3 (compared with 2.5 of casein). The digestibility is 84 %; a very high value compared with other algae, in particular green algae. Toxicity was not detectable in a 90-day test on rats.

Most of the tests were made in the Centraal Instituut voor Voedingsonderzoek, Zeist, Netherlands, and in the Nutritional Division of the Institut Nacional de la Nutrición in México. Active research continues in Mexico and in France, where it is part of the program of the Delegation Générale à la Recherche Scientifique et Technique.

Spirulina has possibilities of being commercialised as poultry food because it is rich in carotenoids, especially xanthophylls which improve the colour of egg yolk. Production in Mexico reached 1000 kg/day in 1974. Prospects for its use as a human food are good because it has been consumed for a long time; and because its price makes it competitive with other foods rich in protein.

In Tchad, since 1947, FAO has made successful nutritional tests. Some tests were also made recently in Mexico. A group of athletes of the Comite Olimpico Mexicano took 20 to 40 g of *Spirulina* daily during 2 periods of 30 to 45 successive days with good results. Tests are now being made on children in Mexico using *Spirulina* merely dried without decolorisation. It can be solvent-extracted if need be. *Spirulina* that has not been heated has an attractive blue-green colour but it becomes dull brown when cooked. (Based on information supplied by Sosa Texcoco SA Mexico.)

6. Green micro-algae

H. TAMIYA

Work on mass-culturing of green micro-algae as a source of inexpensive protein of good quality is proceeding in West Germany, Czechoslovakia, India, Taiwan, Japan, the USA and, probably the USSR. The algae used in Germany and Czechoslovakia are mainly strains of *Scenedesmus*; those employed in Taiwan and Japan are strains of *Chlorella*. In most countries mono-algal cultures are used, but in the USA mixed algal cultures are grown symbiotically with bacteria in sewage disposal ponds. Both in Germany and Czechoslovakia, algal mass-cultures are run under autotrophic conditions using solar radiation as the energy source and supplying CO_2 in the culture solution. In Japan, this type of mass-culturing was discontinued in 1963, when mixotrophic and heterotrophic culture methods were adopted. In the former, the culture solution contains organic material (usually acetic acid) and is exposed to light, and in the latter the culture is run in total darkness (using 'fermenters' as large as 5000 litres) with glucose, acetic acid or ethanol as the carbon and energy source (Takechi, 1971).

The projects of algal mass-culturing in Germany, Czechoslovakia and other countries are more or less academic, but those in Japan and Taiwan are commercial. Thus, for example, the autotrophic culture ponds used in Germany have 170–200 m² of illuminated area (Soeder, 1969; Soeder *et al.*, 1970*a*), while the mixotrophic culture pond now in operation in Okinawa, Japan, has a surface area of 2300 m² and the work is financed by selling the product.

The yield of algae per unit area of autotrophic and mixotrophic cultures naturally depends on the climate. In northern Europe, culture stops during the cold season. In Dortmund, Germany, the dry weight yield of *Scenedesmus* (*S.* 276-3a or *S. acutus* var. *alternans*, Hortobagyi) in summer was reported to be 25 g/m²·day (Soeder *et al.*, 1970*b*). According to the Czechoslovakian scientists (cited in Stengel, 1971), the culture of *Scenedesmus* in the Mediterranean area will give an average yield of 30–2 g/m²·day. The yields (in g/m²·day) obtained by the companies using mixotrophic cultures of *Chlorella* in the Far East are shown in Table 6.1.

The heterotrophic culture of *Chlorella* (*C. regularis* S-50) by a company in Japan gave a yield of 6.5 g (dry weight)/l·day using glucose as

Table 6.1. *Dry matter yields (g/m²·day)*

	Minimum	Maximum	Yearly average
Toyota, Japan (35° N):	1.5 (Jan.)	17.2 (July)	10.3
Taipei, Taiwan (25° N):	3.8 (Jan.)	27.8 (Oct.)	13.0

the carbon source (Takechi, 1971). For obtaining such a yield the culture had to be 'fortified' by the addition of beef- and yeast-extracts and casein hydrolysate.

In general, the dried cell materials of both *Scenedesmus* and *Chlorella* contain 50–65 % protein. The heterotrophically grown *Chlorella regularis* contains 45–65 % protein (Endo & Shirota, 1972). Kanazawa (1964) measured the amino acid composition of bulk protein, peptides and the free fraction from autotrophically (and synchronously) grown *Chlorella ellipsoidea*. Throughout the algal life cycle, the amino acids in protein (per unit dry weight of cells) remained fairly constant, and were twenty to forty times greater than those present as peptides and in the free state. The predominant amino acids in the protein fraction were alanine, glutamic acid, glycine and leucine, and the minor components were histidine, tyrosine, cysteic acid and methionine. This amino acid pattern was, when compared in terms of mole per cent, almost identical with that of the structural protein of chloroplasts isolated from leaves of various species of higher plants which had been analysed in detail by Weber (1959, 1961; cited in Menke, 1962). The same was true of the amino acid spectrum of *Scenedesmus* (Soeder *et al.*, 1970*b*).

Numerous experiments, using rats, mice, pigs, chicken, etc., have been carried out on the digestibility and nutritive value of *Chlorella* and *Scenedesmus*. The former have been carried out mainly in Japan (Takechi, 1971), the latter in West Germany and Czechoslovakia (Soeder *et al.*, 1970*a*; Meffert & Pabst, 1963). The digestibility of the algae depended strongly on the methods of processing, especially that of drying. Japanese workers recommended freeze-drying and spray-drying, while German workers obtained good results with roller-dried algal samples. The appropriately processed algae showed, in general, 75 to 85 % digestibility (Lubitz, 1963; Kraut *et al.*, 1966; Endo & Shirota, 1972); the biological value (BV) was 60 to 78 (Tamura *et al.*, 1957; Mitsuda *et al.*, 1959; Kraut *et al.*, 1966). In all cases, no toxic effects have been noticed.

Experiments with humans were performed in Germany, Japan, the USA and the USSR. According to the Japanese workers the digestibility of dried *Chlorella* cells for adult humans ranged from 75 to 89% (Takechi, 1971). Powell *et al.* (1961) reported that up to 100 g (dry weight) of a mixture of *Chlorella* and *Scenedesmus* per adult man daily were tolerated by all subjects tested. When larger amounts were given, gastrointestinal symptoms – nausea, vomiting, abdominal distension, flatulence, lower abdominal cramping pains, bulky hard stools, etc. – were prominent (see also McDowell & Leveille, 1963). These symptoms, however, disappeared shortly after the algal diet was discontinued. The experiments performed in the USSR (Kondratyev *et al.*, 1966) also showed that 150 g per day per capita caused allergic reactions and negative nitrogen balance.

In contrast, much better results were obtained in Germany with *Scenedesmus* by Kofranyi & Jekat (1967) who gave the alga to adults in gradually increased amounts until finally the whole protein requirement was covered by the alga. An experiment lasting for 3 weeks showed that the biological value (taking that of eggs as 100) of the alga was 81.5 ± 1.5%, which means that 124 g of the algal protein corresponded to 100 g of egg protein. It seems that the *Scenedesmus* produced by German workers is superior to the *Chlorella* grown in Japan not only in its acceptability (taste) and digestibility, but also in its nutritive value. Soeder & Pabst (1970), however, remarked that in using algal cells in human diet one should keep in mind that the high content of nucleic acids in algal cells might cause kidney troubles by formation of uric acid leading to gout. They concluded that the micro-algae should be used as a protein-supplement rather than as the major protein source in human diet.

According to some people, the extract of *Chlorella* is effective in improving the flavour of foods such as soy-sauce, natto, soybean curd, rice-cake, fish-paste, noodles, bread, sausages and whisky (Takechi, 1971). *Chlorella* extracts are also thought to have many pharmacological effects and are regarded by many laymen as panaceas. Consequently, some enterprises are producing *Chlorella* by using expensive heterotrophic or mixotrophic culture methods and selling it for 30 US dollars/kg or more (Takechi, 1971). Thus, the idea of using *Chlorella* as an inexpensive protein source has been totally abandoned in Japan.

Mass-culturing of green micro-algae seems in other countries to be on an authentic track of utilising algal cells as protein sources; although they have not yet reached the stage of 'industrial' production. Oswald & Golueke (1968) estimate that, if a sewage disposal pond of large enough dimensions could be constructed, the price of the mixture of micro-algae and bacteria would be about 8 US cents/kg dry weight.

References

Endo, H. & Shirota, M. (1972). Studies on the heterotrophic growth of *Chlorella* in a mass culture. In: *Fermentation technology today*. Proceedings of the Fourth International Fermentation Symposium, pp. 533–41. Kyoto, Japan.

Kanazawa, T. (1964). Changes of amino acid composition of *Chlorella* cells during their life cycle. *Plant and Cell Physiology*, **5**, 333–54.

Kofranyi, E. & Jekat, F. (1967). Zur Bestimmung der biologischen Wertigkeit von Nahrungsproteinen. XII. Die Mischung von Ei mit Reis, Mais, Soja, Algen. *Zeitschrift für physiologische Chemie*, **348**, 84–8.

Kondratyev, Y. I., Bychkov, V. P., Ushakov, A. S., Boiko, N. N. & Klyushkina, N. S. (1966). Use of 50 and 100 g dry material of unicellular algae in human diets. *Voprosy Pitaniya*, **6**, 9–14.

Kraut, H., Jekat, F. & Pabst, W. (1966). Ausnutzungsgrad und biologischer Wert des Proteins der einzelligen Grünalge *Scenedesmus obliquus*, ermittelt im Ratten-Bilanz-Versuch. *Nutritio et Dieta*, **8**, 130–44.

Lubitz, J. A. (1963). The protein quality, digestibility and composition of alga *Chlorella* 71105. *Journal of Food Science*, **28**, 229.

McDowell, M. E. & Leveille, G. A. (1963). Feeding experiments with algae. *Federation Proceedings*, **22**, 1431–8.

Meffert, M.-E. & Pabst, W. (1963). Über die Verwertbarkeit der Substanz von *Scenedesmus obliquus* als Eiweissquelle in Ratten-Bilanz-Versuchen. *Nutritio et Dieta*, **5**, 235–54.

Menke, W. (1962). Structure and chemistry of plastids. *Annual Review of Plant Physiology*, **13**, 27–44.

Mitsuda, H., Kawai, F., Murakami, K. & Shikauchi, T. (1959). Studies on the use of *Chlorella* as food. III. On the nutritive value of *Chlorella*-protein. *Eiyoo-to-Shokuryo* (*Nutrition and Foods*), **12**, 31–4. (In Japanese.)

Oswald, W. J. & Golueke, C. G. (1968). Harvesting and processing of waste-grown algae. In: *Algae, man and the environment* (ed. Jackson), pp. 371–89. Syracuse University Press, Syracuse, New York.

Powell, R. C., Nevels, E. M. & McDowell, M. E. (1961). Algae feeding in humans. *Journal of Nutrition*, **75**, 7–12.

Soeder, C. J. (1969). Technische Produktion eiweiss-reicher Mikroalgen. *Umschau in Wissenschaft und Technik*, **24**, 801–2.

Soeder, C. J., Hegewald, E., Pabst, W., Payer, H. D., Rolle, I. & Stengel, E. (1970*a*). Zwanzig Jahre angewandte Mikroalgenforschung in Nordrhein-Westfalen. Jahrbuch des Landesamts für Forschung des Landes Nordrhein-Westfalen, pp. 419–45.

Soeder, C. J., Meffert, M.-E., Rolle, I., Pabst, W., Payer, H. D. & Stengel, E. (1970*b*). *Das Dortmunder Verfahren zur Produktion essbarer Mikroalgen.* Kohlenstoffbiologische Forschungsstation e.V., Dortmund, Germany.

Soeder, C. J. & Pabst, W. (1970). Gesichtspunkte für die Verwendung von Mikroalgen in der Ernährung von Mensch und Tier. *Berichte der Deutschen botanischen Gesellschaft*, **83**, 607–25.

Stengel, E. (1971). Anlagentypen und Verfahren der technischen Algenmassen-produktion. *Berichte der Deutschen botanischen Gesellschaft*, **83**, 589–606.

Takechi, Y. (1971). *Chlorella*: Sono-Kiso-to-Ooyo (*Chlorella*: basic knowledge and application). Gakushu-Kenkyu-Sha, Tokyo. (In Japanese.)

Tamura, E., Baba, H., Tamura, T. & Obata, Y. (1957). Nutritional studies on *Chlorella*. VIII. Report of the National Institute of Nutrition, 1957, 9–10. (In Japanese.)

Weber, P. (1959). Artspezifische Unterschiede zwischen lamellaren Strukturproteiden aus Chloroplasten. *Zeitschrift für Naturforschung*, **14**, B, 691–2.

Weber, P. (1961). Über lamellare Strukturproteide aus Chloroplasten verschiedener Pflanzen. Doctoral Thesis, University of Cologne.

PART II

Concentrates made by mechanical extraction

7. Protein products from coconuts

D. A. V. DENDY

The coconut is the fruit of *Cocos nucifera* L., a palm which grows within the tropics, generally on the littoral; an alluvial soil, with high temperature, high rainfall and sunshine being preferred. World production in 1969 was estimated to be 3296500 tonnes of copra and 29765 million nuts (FAO, 1970). As a smallholders' crop the palm provides building materials (palm fronds and tree trunks), fibre goods from the husk (coconut matting, ropes and packing), fuel (shell and shell charcoal), food (protein and oil from the flesh) and drink (the 'water' from inside the kernel). The nuts are available throughout the year and enter international trade as copra, oil, nuts and desiccated coconut: many countries depend on exports of one or more of these as a major source of export earnings.

In plantation, trees may be planted at about 10 m intervals and annual yields of 60 nuts per tree can be expected, though yields of 120 are common in well-run plantations: smallholders may get only 20 to 30 nuts. Fodder and food crops may be grown under the palms and the grazing of cattle is beneficial.

The coconut has been a traditional food in practically all the countries where it is grown and the quantity of fresh coconuts consumed locally varies from over 90 % of the total production, for example in Thailand, to less than 2 % of the total production in the Philippines. Coconut enters into the diet of the people in many ways; in the form of tender nuts used for their water, mature nuts for cooking and the preparation of sweetmeats, and oil for home consumption. Probably the best-known product is coconut milk, the oil/protein/water emulsion obtained when the grated fresh coconut meat (endosperm) is squeezed through a muslin cloth. The consistency of this milk will vary according to whether or not water has been added during the process. In addition, in some places it is the usual practice to repeat the operation two or three times; each time a more dilute emulsion is obtained suitable for different uses in cooking. The efficiency of coconut milk extraction in the home is low and up to 50 % of the oil and protein can remain in the residue.

There have been many attempts to produce a bottled or packed coconut milk, and research is still taking place in, for example, Thailand, the Philippines, Sri Lanka, Fiji and Jamaica. However, so far as is known, no product has achieved commercial success. A commercial method for preparing coconut milk will require a very efficient milling system. This is in fact the first essential stage of a wet process to obtain coconut oil and edible protein, and is the basis of work being carried out at the Tropical Products Institute in London and A and M University, Texas (Samson *et al.*, 1971).

Coconut protein, like most oilseed proteins, is deficient in lysine, methionine and threonine; the ratio of essential amino acids nitrogen (N) to total N is lower than in animal proteins. Coconut protein compares favourably with groundnut protein and improvements in the growth and well-being of under-nourished children when fed with a protein supplement containing coconut meal have been reported (Prasanna *et al.*, 1969).

The chemical composition of the kernel varies considerably (see, for example, FAO, 1958); typical figures are quoted by Krishnamurthy *et al.* (1958). Each dehusked nut contains around 55 % kernel and this is made up of 46.3 % moisture, 37.3 % oil and 4.1 % protein. The protein contained 3.0 g/16 g N of lysine. However, Timmins (1970) found the lysine content of coconut protein isolates to vary from 3.7 to 4.2 g/16 g N. The lysine content would seem to be variable, the following figures (in g/16 g N) having been obtained by various workers: 3.0 for whole flesh (Krishnamurthy *et al.*, 1958); 2.2 for cake (Sreenivasan, 1967); 4.2 for protein isolate (Timmins, 1970); 3.3 for meal (Samson *et al.*, 1971); 2.3, 2.5, 2.2, 3.1 and 3.0 for meal (Altschul, 1958).

The biological value (BV) of copra meal is 58 and protein efficiency ratio (PER) 1.2 (see also Sreenivasan, 1967) but extracted proteins are said to have much better quality – for example, Rama Rao *et al.* (1964) found a PER of 1.88 for solvent-extracted meal, and 1.86 for a coconut protein concentrate.

Up to now the market economy for coconuts has been based on the need for coconut oil. It has been calculated that for every 1 kg of coconut protein recovered, 7 kg of coconut oil must be processed. Therefore, the total emphasis has been on maximum oil recovery at the lowest possible cost. The cheapest method is, it would seem, to dry the flesh to give copra from which the oil is then expelled. Copra processing is often insanitary and the copra meal itself suitable only for animal feed: if protein fit for human consumption could be obtained, the coconut would prove a potentially valuable source. Improved variants of the copra process have not, generally, proved successful or economic, partly because the product is still essentially copra meal; that is, it contains cellulose as well as protein and is not acceptable as human food.

The fresh or 'wet' processes which use fresh *undried* coconut meats for immediate processing into oil and protein flour continue to challenge the imagination because they offer the potential for obtaining coconut protein products as well as coconut oil of superior quality (FAO, 1968; Timmins, 1970; TPI, 1973). It must be borne in mind, however, that if copra is correctly dried and processed, oil of remarkably good quality can be obtained.

Two major operations are involved in wet processes:

1. The separation of an emulsion of oil, protein, water etc., from the fibrous part of the kernel. As far as yield of oil is concerned, it is at this stage that most if not all wet processes tried so far have failed, in that too large a proportion of the emulsion is left with the cellulose. A milling or pressing operation is a feature of this stage.

2. The splitting of the emulsion, or milk, into oil, protein and an aqueous phase, has proved a very difficult operation to carry out continuously, cheaply and so efficiently that clear oil is obtained in yields of approaching 100 %. In most processes a centrifuge is used to split the emulsion, which is usually pretreated by heat, enzymes, acids etc.

All wet processes require as starting material fresh kernels and these are not always easy to obtain, even for a pilot plant.

Whereas copra may be prepared in villages and small factories, and can then be transported without deterioration (provided the moisture content is less than 6 %), fresh coconut meat must be used almost as soon as it is extracted from the shell. The wet process plant must, therefore, be sited where there is an adequate supply of fresh coconuts plus the labour needed to obtain the meat. Furthermore most wet processes require adequate supplies of water, electricity and the means of disposing of aqueous as well as solid wastes. The unwise siting of pilot plants may have led to premature abandonment of otherwise promising processes.

None of the wet processes is in commercial use, possibly because the apparent yield of oil is low: that is, whereas the yield of oil from copra is over 95 %, that from fresh coconuts is rarely above 80 %. However, if one considers oil and other losses during drying, storage and shipment of copra, the overall oil yield from tree to refined oil is approximately the same for wet processes as for the copra process. Indeed, unless copra is dried by modern, hygienic methods the losses can be very high due to degradation of the copra by mould. Another reason for using wet processes is that the by-product, edible protein, is a valuable food in coconut-producing countries where protein malnutrition is endemic. It is therefore hoped that research will continue in order to find a cheap and effective means of extracting oil and edible protein from coconut.

References

Altschul, A. M. (1958). *Processed plant protein foodstuffs*. Academic Press, New York.

Block & Weiss (1956). *Amino acid handbook*. Thomas, Springfield, Illinois.

FAO (1958). Copra processing in rural industries. Agricultural Development Paper No. 63. FAO, Rome.

FAO (1968). Coconut oil processing. Agricultural Development Paper No. 89. FAO, Rome.

FAO (1970). *Production Yearbook 24*. FAO, Rome.

Krishnamurthy *et al.* (1958). The chemical composition and nutritive value of coconut and its products. *Food Science*, **7**, 365–9.

Prasanna, H. A., Rama Rao, G., Desai, B. L. & Chandrasekhara, M. R. (1969). Use of a spray-dried infant food based on coconut in the treatment of protein malnutrition (kwashiorkor). *Journal of Food Science and Technology*, **6**, 187–8.

Rama Rao, G., Indira, K., Bhima Rao, U. S. & Ramaswamy, K. G. (1964). Protein efficiency ratio of coconut flour and some products from it produced by azeotropic process. *Journal of Food Science and Technology*, **1**, 23–5.

Samson, A. S., Cater, C. M. & Mattil, K. F. (1971). Preparation and characterization of coconut protein isolates. *Cereal Chemistry*, **48**, 182–90.

Sreenivasan, A. (1967). The use of coconut preparations as a protein supplement in child feeding. *Journal of Food Science and Technology*, **4**, 59–65.

Timmins, W. H. (1970). A process for the extraction of oil and protein from fresh coconuts. M. Tech. Thesis. University of Loughborough, England.

TPI (1973). Tropical Products Institute Report on the wet processing of coconuts. G78, Technology; G79, Economics. Tropical Products Institute, London.

8. Soybeans: processing and products

S. J. CIRCLE & A. K. SMITH

Although the grain crops of the world contribute almost half of the total protein of the world food supply, these grain proteins are deficient in varying degree in several of the essential amino acids, namely lysine, methionine, threonine and tryptophan. In countries lacking adequate animal protein sources, this amino acid imbalance fortunately can be appreciably improved by mutual supplementation with properly processed oilseeds, pulses and other legumes, and nuts (Liener, 1972). Soybeans now supply more protein than any other crop except wheat, maize and rice. However, most of the soybean crop is processed by solvent extraction to yield edible oil, with the meal primarily directed to animal feeding; only a minor percentage of the meal is processed directly for human consumption. Thus in the United States only about 1.5 % of the defatted soy meal is used directly in the human food supply (Circle & Smith, 1972); in China, Japan and other far eastern countries the proportion is greater.

For several years the United States soybean crop has amounted to about three-quarters of the world soybean production. The US together with China and Brazil accounted for 93.5 % of the world soybean crop in 1972. Table 8.1 (USDA, 1974) summarizes world soybean production by individual countries for the years 1967–72. The 1972 US crop was 34.9 megatonnes (Mt), and in October 1973 was estimated to be 43.5 Mt, assuming average yields and weather conditions. Approximately one-half of the US crop was exported in the form of seed, meal and oil. In June 1973 the price of soybeans in the US market soared briefly to $478/t, almost four times the autumn 1972 price of $123. This phenomenal advance in soybean prices was due to unprecedented demand sparked by adverse weather conditions worldwide and by the disastrous shortfall of the Peruvian anchoveta catch, which precipitated a shortage of animal feeds resulting in severe restriction of meat, milk, poultry and egg production. However, the price of soybeans became more reasonable with harvest of the 1973 crop (cash price on October 2, 1973 was $218/t).

Even though a major portion of the world's soybean supplies is

Table 8.1. *Soybeans: production in specified countries and the world 1967–73*

Continent and country	Production (tonnes × 10^3) 1967	1968	1969	1970	1971	1972[a]	1973[a]
North America							
United States[b]	26575	30127	30839	30675	32006	34916	42634
Canada	220	246	209	283	280	375	397
Mexico	121	270	300	240	250	375	510
South America							
Argentina	20	22	32	27	59	78	272
Brazil	716	654	1057	1609	2100	3340	5000
Colombia	80	87	100	95	100	115	97
Paraguay	18	14	45	52	75	115	122
Europe							
Romania	41	47	51	91	165	186	244
Yugoslavia	9	3	5	5	4	6	5
USSR	543	528	434	603	535	358	423
Africa							
Nigeria[c]	16	7	34	11	1	4	1
Tanzania[d]	4	4	4	4	4	4	4
South Africa[e]	4	5	7	4	2	3	5
Asia							
Iran	2	2	4	6	7	9	20
Turkey	6	8	11	12	13	13	13
China							
Mainland	6950	6480	6300	6900	6700	6300	6700
Taiwan	75	73	67	65	61	60	60
Cambodia (Khmer, Republic of)	4	4	4	4	4	4	4
Indonesia	416	420	389	488	475	518	539
Japan	190	168	136	126	122	127	118
Korea, South	201	245	229	232	222	225	246
Philippines	1	1	1	1	1	1	2
Thailand	53	45	61	70	74	83	95
Australia	1	2	2	6	11	26	38
Other countries	273	278	282	296	309	332	351
Total excluding Romania, USSR, Bulgaria, Hungary, Mainland China, North Korea and North Vietnam[f]	28752	32424	33557	33961	35905	40456	50285
Estimated World total[f]	36543	39740	40508	41810	43574	47677	57903

Years shown refer to years of harvest. Southern Hemisphere crops which are harvested in the early part of the year are combined with those of the Northern Hemisphere harvested the latter part of the same year. [a] Preliminary, [b] harvested for beans, [c] quantities purchased by the Nigerian Marketing Boards for export, [d] sales, [e] European farms only, [f] includes estimates for the above countries for which data are not available and for minor producing countries.

USDA (1974). Foreign Agriculture Service. Prepared or estimated on the basis of official statistics of foreign governments, other foreign source materials, reports of US Agricultural Attaches and Foreign Service Officers, results of office research and related information.

currently processed by defatting with solvents (usually hexane), this approach requires heavy capital investment (De, 1971), which would impose an inordinate or impossible burden on most of the developing countries. In this chapter emphasis will be placed on methods of processing soybeans which avoid the use of solvents.

Agronomy

The soybean, *Glycine max* (L.) Merrill (a native of East Asia), is a member of the family Leguminosae, subfamily Papilionaceae (Godin & Spensley, 1971; Howell & Caldwell, 1972). Several other nomenclatures have been reported: *Glycine hispida* and *Soja max* (Piper & Morse, 1923); *Phaseolus max* L., *Dolichos soja* L., *Soja hispida* Moench, *Soja japonica* Savi, *Glycine soja* Sieb & Zucc., and others (Williams, 1950). Although essentially subtropical in origin, cultivation extends from tropical regions up to latitude 52° N. Selection of the appropriate variety is important to fit the agronomic conditions. Genetic, morphological, cultural, and other biological characteristics are given comprehensive discussion in the foregoing references.

Varieties

Present varieties in the United States (Anon., 1972) primarily stem from the introduction early this century of Oriental strains, followed by breeding and selection under the direction of the US Regional Soybean Laboratory in Urbana, Illinois, which develops new high-yielding, disease resistant strains, and also maintains a soybean germplasm bank. New varieties are turned over to commercial seedsmen for the replacement of older strains. Currently recommended varieties in the United States are listed by Howell & Caldwell (1972), Cooper (1971), and in Anon. (1969, 1972).

Cultivation

The soybean is known as a short-day plant, day-length being the primary factor in the change from the vegetative to the reproductive stage. A given variety is generally suited to a relatively narrow band of latitude ranging from 160 to 240 km wide.

The interval elapsing between sowing and harvest varies widely, depending on variety, latitude, soil and weather conditions. In the United States corn belt, the time to reach the green moist edible stage for vegetable (garden) type soybeans is 90 to 120 days after planting. Mature dry field (commercial) type soybeans, need 126–39 days in the northern states and 136–58 days in the southern states (Anon., 1969).

In introducing soybean production into a new area, several important factors must be observed (Scott & Aldrich, 1970; Howell & Caldwell, 1972): (*a*) varieties must be selected which grow to maturity during the growing season at the given latitude; (*b*) the seed should be inoculated with the symbiotic nitrogen-fixing bacteria *Rhizobium japonicum* (Kirchner) Buchanan; (*c*) if the pH of the soil is below 6, it should be treated with lime to bring it to pH 6–6.5; (*d*) proper selection of fertilizers to achieve optimum yields is still under investigation, but replacement of potassium and phosphorus in the soil is necessary and frequently other minerals are added (Scott & Aldrich, 1970).

The FAO has published an article which discusses research and practice in introduction of the soybean for cultivation in various nations of Asia, Africa, and Latin America (FAO, 1967; see also De & Russell, 1967).

Yields

Soybean yields vary widely; the average in the US for 1972 was 1880 kg/ha. However, some farmers have obtained yields as high as 2700–3000 kg/ha; indeed, in some production contests, yields have been reported in the range 5400–6700 kg/ha, and in special experimental plots even higher. The US average of 1880 kg/ha for yield of soybeans translates into 120 kg/ha of nitrogenous matter or 750 kg/ha of protein (based on a nitrogen (N) content of 6.4 % and N conversion factor of 6.25). Experimental growth regulators have been reported to show some promise of enhancing soybean yields (Anon., 1973).

Soybean composition

Soybean varieties vary in color, shape, size, and composition of their seed. This subject has been reviewed in detail by Smith & Circle (1972*a*), including reference to seed parts, protein and non-protein N, amino acid composition, oil composition, mineral constituents and carbohydrates. The protein content of the whole seed, approximately 40 % (% N × 6.25) of the seed weight, is the highest of any commercially important crop. The non-protein N calculated as a percentage of total N ranges from a low of 2.9 % to a high of 7.8 %. The seed coat weight is approximately 8 % of the seed weight, but varies with the size of the seed. The N content of the seed coat is about 1.5 %.

Because of the increasing demand for protein in the US and abroad, US plant breeders are presently developing new higher protein content varieties (not yet commercial) in the range 45–50 % (% N × 6.25). Usually with an increase in protein, there is a decrease in oil, approximately 0.5 % oil loss for each 1 % protein gain.

Protein nutritional value

The soybean contains several trypsin inhibitors (Rackis, 1972) which reduce the nutritive value of the protein for rats and other animals when beans are fed in the raw state; however, 10–15 minutes of moist heat treatment at 100 °C inactivates these inhibitors, and is the most common type of processing given to full-fat and defatted soy meal products for feed and food uses. This same treatment diminishes the so-called 'beany' and 'bitter' flavors. A study of heat inactivation of trypsin inhibitor, lipoxygenase and urease in soybean meats with the use of acid and base additives was reported by Baker & Mustakas (1973). In soy protein isolates and concentrates, the trypsin inhibitors are reduced to innocuous levels.

Liener (1972) has reviewed the nutritional value of soybean protein in the whole bean and as it occurs in the various processed fractions, including immature green and mature beans, sprouts, meal and flour (full-fat and defatted), protein concentrates, protein isolates, tofu curd, soy fermentation products, soybean milk and textured meat analogs. Based on the ratio of total essential amino acids to total N, soy flour ranks just below the animal proteins (egg, milk, meat and fish), and above the other oilseeds (sesame, cottonseed and peanut). Soy protein isolate is somewhat lower in nutritional value than the protein in soy flour, due to the separation of soy 'whey' proteins in processing.

The limiting amino acids on the basis of chemical score are the sulfur-containing methionine and cystine; however, one of the most valuable attributes of soy protein lies in the fact that it contains more lysine than most plant proteins (Liener, 1972).

Robinson *et al.* (1971) emphasized the complementarity of the amino acids of soy milk with those of the cereal grains, using the essential amino acid index (EAAI) as criterion; they also found that most varieties had PER values of 2.1 compared with the casein standard of 2.5; methionine supplementation with proper heat processing raised the PER almost to that of casein.

Traditional processing into nonfermented foods

The principal nonfermented soy foods processed traditionally include the table vegetable, soy milk, tofu (soybean curd), yuba, kinako and soybean sprouts (Piper & Morse, 1923; Smith, 1958; Watanabe, 1969; Smith & Circle, 1972*a*, *b*).

Soybeans as a table vegetable

It is regrettable that the use of soybeans as a table vegetable is relatively neglected, considering that this mode of use requires the least processing. Green soybeans are rarely available in the market; although similar in flavor and texture to lima bean, consumption of cooked green soybeans is apparently confined to rural areas, to home gardeners and to health food advocates. Certain large-seeded varieties are often referred to as 'garden' or 'vegetable' type soybeans in contrast to the better known (smaller seeded) 'field' type. These garden varieties are not suitable for commercial oilseed production because of lower yields and harvesting problems, such as shattering from the pods when mature. They are suitable in the green stage for preservation in the frozen, canned or dehydrated condition (Piper & Morse, 1923; Smith & Van Duyne, 1951), and in the mature stage for cooking or canning. Woodruff & Klaas (1938), Lloyd & Burlison (1939) and Lloyd (1940) investigated a number of varieties of vegetable type soybeans, and compared the cooked beans for flavor, color, texture and other qualities.

Breaded, deep-fat fried vegetable soybeans, made from beans harvested in the green stage, were recommended as a potential high-protein snack item (Collins & Ruch, 1969).

Soy milk

In the far east, soy milk is traditionally prepared by soaking the beans in water overnight, draining the water, grinding the beans in hot water (approximately 10 parts water to 1 part beans to give the desired solids content in the final product), then cooking the wet mash for 10–15 minutes at 100 °C, and filtering through coarse cloth to recover the product. The water–bean ratio specified above extracts about 50 % of the protein and oil.

The characteristic flavor when made in this manner may limit acceptance. Several references (Kon *et al.*, 1970; Robinson *et al.*, 1971; Circle & Smith, 1972; Baker & Mustakas, 1973) indicate that lipoxidase and perhaps other enzymes in the soybean may be responsible for much of this 'off-flavor'; the activity of these enzymes can be diminished by grinding the beans initially in hot water at 80–100 °C (with or without chemical additives in the water).

When supplemented with vitamins, minerals, methionine, oil and sugar, soy milk is nutritionally equivalent to cow's milk (Shih, 1970; Liener, 1972; Badenhop & Hackler, 1973). In Japan spray-dried soybean milk is made commercially (Watanabe, 1969). Soy milk may be used to 'tone' or supplement buffalo milk or cow's milk (PAG, 1973).

Tofu (soybean curd)

Tofu, which is a good source of protein and oil, is the most important of the non-fermented foods. The first step is to prepare soy milk as described above. While stirring the hot milk, the tofu is precipitated by adding calcium sulfate to form a gelatinous curd, which is separated and drained free of whey, and then soaked in fresh water for an hour or more; it is then ready for the market (Smith, 1958; Smith *et al.*, 1960; Watanabe, 1969; Smith & Circle, 1972*b*). In 1957 Japan had 40000 tofu plants registered in their Tofu Association (Smith, 1958), and there were many more small family plants. The yield of tofu is variable, depending of the variety of beans, as well as other factors. Smith *et al.* (1960) reported that 1.8 kg of soybeans (12 % moisture) yielded 5.55 kg of tofu (88 % water). On a dry basis, this is 42 % of the bean weight. Dry tofu contains 55 % protein and 28 % oil.

Tofu curd (wet) is frequently served in soup, and may also be used in the dry form, called kori-tofu (Watanabe, 1969). Another popular way to serve tofu is to deep-fat fry the wet curd to give the product named aburage in Japan. In the Philippines, soybean curd is known as tokua and taku (Robinson *et al.*, 1971).

Yuba

When soy milk is cooked at or near its boiling temperature a film composed of protein and oil (with a composition similar to tofu) forms on the surface. This film when removed and dried is a nutritious food known as yuba. It is used in soups or fried in fat for table use (Watanabe, 1969; Smith & Circle, 1972*b*).

Concentrates by mechanical extraction

Kinako

A product made in Japan known as kinako is similar to full-fat soy flour except that it may contain the seed coat (Watanabe, 1969). Kinako is prepared from clean whole soybeans which are first roasted in an oven at about 100 °C for 30 minutes to diminish the beany flavor and improve the nutritional value and then ground into a fine powder. This product is sprinkled on rice or rice cakes and used as a protein supplement.

Salted soybeans

A product resembling salted peanuts is prepared by soaking clean whole beans in a 10 % salt solution overnight, then boiling in water for about 30 minutes. The beans are removed from the water and dry-roasted to a light brown color. The yellow seed coat varieties are preferred for their better appearance (Piper & Morse, 1923).

Soybean sprouts

When properly prepared, soybean sprouts do not have the beany flavor so frequently associated with soy foods. Chen (1962), Piper & Morse (1923), Smith & Circle (1972*b*) and Smith & Van Duyne (1951) have described some of the common procedures for their preparation. Clean whole beans are soaked in water overnight and sprouted in glass fruit jars, flower pots or between wet cloths supported by a wire screen. They are maintained in the dark during the sprouting period and washed several times a day with water containing one teaspoon of calcium hypochlorite per 13 litres of water to discourage the growth of micro-organisms. Excessive moisture is unfavorable to rapid sprouting; the recommended amount is about 55 % of the weight of the wet beans at a temperature of about 30 °C. The sprouts should be about 5 cm long in 4–5 days.

The sprouts may be prepared for the table by pan frying with a small amount of fat, and seasoning with salt. For use in a salad they are boiled in water for a few minutes, drained and chilled. The sprouted beans are rich in protein and oil and during sprouting there is a rapid increase in ascorbic acid as well as other vitamins.

Traditional processing into fermented foods

Foods that are processed by fermentation have been used in the Orient since ancient times; they contribute flavor as well as protein to the diet.

Although many different foods are included in this category the principal ones are miso, shoyu and tempeh; others include natto, hamanatto, sufu (soy cheese), tao-tjo, kochu chang, ketjap, ontjom, and yogurt-like products (Piper & Morse, 1923; Smith, 1949; Watanabe, 1969; ISFM, 1972; Smith & Circle, 1972*b*).

Miso and shoyu

Miso is made by fermenting soybeans and rice, whereas shoyu is made by fermenting soybeans and roasted wheat or wheat flour. The fermenting organisms for both products have been reported to belong to the genus *Aspergillus*, although other organisms and enzymes enter into the operation. Shoyu or soy sauce is a well known condiment that contributes flavor rather than nutrition to the diet but is of special importance as a flavor ingredient for people who depend largely on vegetable sources for their protein.

Miso is used mostly as a soup base at levels high enough to contribute nutrition in the form of amino acids as well as flavor. In some areas of the Orient these products are still processed in the home or as cottage industries; however in recent years in Japan they have developed the fermentation processes to the highest level of sophistication. The ratio of soybeans, rice and salt in making miso is variable but on a dry basis will approximate 100:50:45.

The preparation of miso is described by Ebine (in PISFM, 1972, pp. 127–32), Hesseltine & Wang (1972) and Watanabe (1969). Some of the small scale cottage industries for processing these products are described by Smith (1949). Shoyu processing is discussed by Yokotsuka (in ISFM, 1972, pp. 117–25) as well as by Hesseltine & Wang (1972) and Watanabe (1969).

Tempeh

Tempeh is made in Indonesia by fermenting soybeans with *Rhizopus*. When fried in oil it has a pleasant flavor, aroma and texture and is readily acceptable to Western as well as Oriental tastes (Hesseltine & Wang, 1972; and Santon in PISFM, 1972, pp. 133–9). In Indonesia tempeh is made at home and is usually eaten the day it is made; recently, methods have been studied for extending its shelf life (Hesseltine & Wang, 1972).

In the traditional method, the beans are soaked in water overnight, the seed coat removed by hand, and the beans boiled in water for

30 minutes, drained and spread on a cloth for surface drying. The beans are inoculated with small pieces of tempeh from a previous preparation and allowed to ferment at room temperature for a day. By that time the beans are bound together with a white mycelium. The cake is cut into thin slices, dipped into a salt solution and fried in coconut oil. Also it may be baked or added to soup.

Contemporary processing without defatting

'Debittering' by aqueous treatment

In the past fifty years or more, considerable effort has been expended on developing so-called 'debittering' methods for improving the flavor of soybean products. Some of these employ dry heat, but most treat whole, flaked, or comminuted full-fat soybeans with aqueous solutions in combination with moist heat; the debittered full-fat products may then be subjected to defatting.

A good deal of the undesirable volatile components can be removed by dry heat in pressure or vacuum ovens, but it is more effective to use moist heat by application of steam in various types of equipment; the latter approach is the most widely used at present. Nonvolatile components are removed by immersion in a liquid into which they can diffuse. Deep-fat frying is one approach, but the more commonly advocated immersion methods rely on water or aqueous solutions employing a host of assisting chemicals both inorganic and organic. Enzymic treatment and fermentation with micro-organisms have also been advocated.

Thus in a patent review (compiled by Smith, 1945; see also Burnett, 1951; Noyes, 1969; Maga, 1971; Circle & Smith, 1972; Cowan *et al.*, 1973) the methods described include (along with heat) soaking with various alkaline reagents such as sodium, potassium and ammonium hydroxides, carbonates and bicarbonates, and lime; various acids such as hydrochloric, acetic, phosphoric, citric, sulfurous and sulfuric, or salts thereof (sodium, calcium, magnesium, or iron); various organic compounds including alcohols, aldehydes, ketones, amines, esters, ethers; oxidizing agents (hydrogen peroxide, chlorine, or ozone); activated carbon treatment; and treatment with gases (CO_2, SO_2, ethylene, or N_2). Several of these treatments may result in some damage to certain amino acids, such as cysteine, methionine and lysine, and should be used with caution.

Most of this 'debittering' work is difficult to assess for effectiveness without actually carrying out the described processes, and evaluating by flavor-panel testing (Kalbrener *et al.*, 1971; Smith & Circle, 1972*b*). A number of soybean treatment approaches are selected for discussion below, with emphasis on those not relying on organic solvents. However, it has been pointed out (Circle & Smith, 1972), that the most effective treatment of flaked soybeans for improving flavor and color appears to be extraction with ethanol, keeping temperatures as low as is practical during processing.

Whole bean processing

Dry-roasting of whole beans or dehulled meats is the least complicated type of processing (Guidarelli *et al.*, 1964; Guidarelli, 1968; Badenhop & Hackler, 1971; Borchers *et al.*, 1972). Nutritional value of roasted soybeans was inversely related to the degree of roast (Robinson *et al.*, 1971). Deep-fat frying was employed following soaking in water (McComb, 1937) or brine (Moulton, 1938; Nohe, 1938). Soybeans were cooked by aqueous immersion, then dehydrated (Hand *et al.*, 1964; Nelson *et al.*, 1971). Similar cooking was conducted with additives in the water such as alkaline salts (Rockland, 1972; Shemer *et al.*, 1973). Pichel & Weiss (1967) prepared a nut-butter from soybeans analogous to peanut butter.

Full-fat flour

Extrusion cooking to produce full-fat soybean flour has received considerable attention (Albrecht *et al.*, 1967; Bookwalter *et al.*, 1971; Conway, 1971; Mustakas, 1971; Mustakas *et al.*, 1971, 1972; Circle & Smith, 1972; Fox, 1972). The more conventional approach of non-extrusion steam cooking has also been reviewed (Learmonth & Wood, 1962; Schlosser & Dawson, 1969; Circle & Smith, 1972).

Soy milk and curd

Several recent reports refer to improvements in soy milk processing (Bourne, 1970; De, 1971; Robinson *et al.*, 1971; Smith & Circle, 1972*b*; PAG, 1973). The effect of processing conditions was studied in relation to yield, composition, removal of oligosaccharides, control of oxidative 'off-flavors', and nutrient availability (Lo *et al.*, 1968; Wilkens & Hackler, 1969; Sugimoto & Van Buren, 1970; Kon *et al.*, 1970, 1971). Other papers on soy milk and beverage-base processing include

Concentrates by mechanical extraction

Badenhop & Hackler (1973), Mustakas *et al.* (1971, 1972) and Shih (1970). Schroder & Jackson (1971, 1972) reported on efforts to prepare soybean cheese and curd with reduced beany flavor.

Contemporary defatting processes

Defatting by aqueous processing

Although several processes have been described for separating oil by centrifugation of aqueous extracts of comminuted oilseeds, part of the oil emulsifies with the protein to form products containing variable amounts of oil (Circle & Smith, 1972). Rhee *et al.* (1973) claim that, with peanuts, the formation of protein–lipid complexes can be avoided by comminuting the oilseed at pH value in the isoelectric range of the protein.

Defatting with organic solvents

Most of the world soybean crop is processed with hexane to extract the oil (Norris, 1964) and produce defatted meal which is the base for defatted soy flour, protein concentrates, protein isolates, modified isolates, and textured protein products (Circle & Smith, 1972). Solvent-type processing plants range in capacity from 50 to 2000 t or more daily. A worldwide listing of solvent-type soybean processing plants is available in the Soybean Blue Book (Anon., 1972). De (1971) has given details and estimated costs for solvent plants of daily capacities 30, 50, or 100 t, and also discusses production size and costs of pilot and larger plants for full-fat, defatted and refatted soy flours, soy protein isolates, soy milk and tempeh.

Composite flour

Substitutes or extenders for wheat flour in baked goods and pasta products are termed 'composite flours', whether or not they contain some wheat. Many of these are based on mixtures of soy protein products with starches derived from cassava or other roots, or from maize or other non-wheat grains (see Dendy *et al.*, 1972 for a recent bibliography). A supplementary food mixture (CSM) is described (Combs, 1967).

Soy flours, protein concentrates, protein isolates and textured protein products

Much attention in the past 35 years has been devoted to protein fractionation after hexane defatting of the soy meal. References are: Orr &

Adair (1967), Noyes (1969), Ashton *et al.* (1970), BFMIRA (1970), Burke (1971), De (1971), Circle & Smith (1972), Orr (1972), Smith & Circle (1972*b*), van den Burg (1972) and Gutcho (1973*a*, *b*); see also Anon. (1966).

Recipes for using soy protein products in foods are available in several reports (Piper & Morse, 1923; Whiteman & Keyt, 1938; Weiss *et al.*, 1942; Bowman *et al.*, 1945; Van Duyne, 1950; Circle & Johnson, 1958; Chen, 1962; Scholsser & Dawson, 1969).

References

Albrecht, W. J., Mustakas, G. C., McGhee, J. E. & Griffin, E. L. Jr (1967). A simple method for making full-fat soy flour. *Cereal Science Today*, **12**, (No. 3), 81–3.

Anonymous (1966). Soybean processing and utilization. A selected list of references 1955–1965. Library List No. 83. National Agricultural Library, USDA, Washington, DC.

Anonymous (1969). *Soybean farming*. National Soybean Processors Association, Washington, DC, and National Soybean Crop Improvement Council, Urbana, Illinois.

Anonymous (1972). Blue Book Issue. *Soybean Digest* **32** (No. 6). American Soybean Association, Hudson, Iowa. (Prior and 1973 editions also available.)

Anonymous (1973). Growth regulators – a new boom coming? *Soybean Digest*, **33** (No. 9), 10–11.

Ashton, M. K., Burke, C. S. & Holmes, A. W. (1970). *Textured vegetable proteins I*. Science & Technology Surveys No. 62. British Food Manufacturing Industries Research Association, Leatherhead, Surrey, UK.

Badenhop, A. F. & Hackler, L. R. (1971). Protein quality of dry roasted soybeans: amino acid composition and protein efficiency ratio. *Journal of Food Science*, **36**, 1–4.

Badenhop, A. F. & Hackler, L. R. (1973). Methionine supplementation of soy milk to correct cystine loss resulting from an alkaline soaking procedure. *Journal of Food Science*, **38**, 471–3.

Baker, E. C. & Mustakas, G. C. (1973). Heat inactivation of trypsin inhibitor, lipoxygenase and urease in soybeans: effect of acid and base additives. *Journal of the American Oil Chemists' Society*, **50**, 137–41.

BFMIRA (1970). *Protein symposium*. 3. *Vegetable proteins*. Symposium Proceedings No. 8, December. British Food Manufacturing Industries Research Association, Leatherhead, Surrey, UK.

Bookwalter, G. N., Kwolek, W. F., Black, L. T. & Griffin, E. L. Jr (1971). Corn meal–soy flour blends: characteristics and food applications. *Journal of Food Science*, **36**, 1026–32.

Borchers, R., Manage, L. D., Nelson, S. O. & Stetson, L. E. (1972). Rapid improvement in nutritional quality of soybeans by dielectric heating. *Journal of Food Science*, **37**, 333–4.

Bourne, M. C. (1970). Recent advances in soybean milk processing technology. PAG Bulletin No. 10, pp. 14–22. Protein Advisory Group of the United Nations, United Nations, New York.

Bowman, F., Maharge, L., Mangel, M. & McDivitt, M. (1945). Culinary preparation and use of soybeans and soybean flour. University of Missouri Agriculture Experimental Station Bulletin, No. 485.

Burke, C. S. (1971). *Textured vegetable proteins II.* Science and Technology Surveys No. 68. British food Manufacturing Industries Research Association, Leatherhead, Surrey, UK.

Burnett, R. S. (1951). Soybean protein food products. In: *Soybeans and soybean products* vol. 2 (ed. K. S. Markley), pp. 949–1002. Wiley-Interscience, New York.

Chen, P. S. (1962). *Soybeans for health, longevity and economy*. The Chemical Elements, South Lancaster, Massachusetts.

Circle, S. J. & Johnson, D. W. (1958). Edible isolated soybean protein. In: *Processed plant protein foodstuffs* (ed. A. M. Altschul), pp. 399–418. Academic Press, New York.

Circle, S. J. & Smith, A. K. (1973). Processing soy flours, protein concentrate and protein isolates. In *Soybeans: chemistry & technology*, vol. 1, *Proteins* (ed. A. K. Smith & S. J. Circle), pp. 294–338. Avi Publishing Co., Westport, Connecticut.

Collins, J. L. & Ruch, B. C. (1969). Breaded, deep-fried vegetable soybeans. *Food Product Development*, **3** (No. 6), 40–1.

Combs, G. F. (1967). Development of a supplementary food mixture (CSM) for children. PAG Bulletin No. 7, pp. 15–24. Protein Advisory Group of the United Nations, United Nations, New York.

Conway, H. F. (1971). Extrusion cooking of cereals and soybeans. *Food Product Development*, **5** (No. 2), 27–31; **5** (No. 3), 14–22.

Cooper, R. L. (1971). Geographical distribution of soybeans and soybean varieties. In: *Soy: the wonder bean*, pp. 1–7. Symposium, Southern California Section AACC, Aug. 19–21. American Association of Cereal Chemists, St Paul, Minnesota.

Cowan, J. C., Rackis, J. J. & Wolf, W. J. (1973). Soybean protein flavour components: a review. *Journal of the American Oil Chemists' Society*, **50**, 426A.

De, S. S. (1971). *Technology of production of edible flours and protein products from soybean.* FAO, Rome.

De, S. S. & Russell, J. S. (1967). Soybean acceptability and consumer adoptability in relation to food habits in different parts of the world. In: *Soybean protein foods* (ARS-71-35), pp. 20–7. Agricultural Research Service, US Dept. of Agriculture, Washington, DC.

Dendy, D. A. V., James, A. W. & Clarke, P. A. (1972). *Composite flour technology bibliography*. Report G71. Tropical Products Institute, London.

FAO, Nutrition Division, (1967). Soybean; production, cultivation, economics of supply, processing and marketing. PAG Bulletin No. 7, pp. 24–44. Protein Advisory Group of the United Nations, United Nations, New York.

Fox, W. F. (1972). Method of treating soybeans. US Patent 3695891 (October 3).

Godin, V. J. & Spensley, P. C. (1971). In: *Oil and oilseeds*, TPI Crop and Products Digest No. 1, pp. 143–51. Tropical Products Institute, London.

Guidarelli, E. J. (1968). Soybean treating process. US Patent 3407073 (October 22).

Guidarelli, E. J., Eversole, R. A. & Lawrence, J. F. (1964). Treatment of soybeans. US Patent 3141777 (July 21).

Gutcho, M. (1973*a*). *Textured foods and allied products*. Food Technology Review No. 1, Noyes Data Corp., Park Ridge, NJ.

Gutcho, M. (1973*b*). *Prepared snack foods*. Food Technology Review No. 2. Noyes Data Corp., Park Ridge, NJ.

Hand, D. B., Steinkraus, K. H. & Gage, M. H. (1964). A new soybean food, precooked dehydrated soybeans. *Soybean Digest*, **24** (No. 8), 28–9.

Hesseltine, C. W. & Wang, H. L. (1972). Fermented soybean food products. In *Soybeans: chemistry & technology*, vol. 1, *Proteins* (ed. A. K. Smith & S. J. Circle), pp. 389–419. Avi Publishing Co., Westport, Connecticut.

Howell, R. W. & Caldwell, B. E. (1972). Genetic and other biological characteristics. In *Soybeans: chemistry & technology*, vol. 1, *Proteins* (ed. A. K. Smith & S. J. Circle), pp. 27–60. Avi Publishing Co., Westport, Connecticut.

Kalbrener, J. E., Eldridge, A. C., Moser, H. A. & Wolf, W. J. (1971). Sensory evaluation of commercial soy flours, concentrates and isolates. *Cereal Chemistry*, **48**, 595–600.

Kon, S., Wagner, J. P., Becker, R., Booth, A. N. & Robbins, D. J. (1971). Optimum nutrient availability of legume food products. *Journal of Food Science*, **36**, 635–9.

Kon, S., Wagner, J. P., Guadagni, D. G. & Horvat, R. J. (1970). pH adjustment control of oxidative off-flavors during grinding of raw legume seeds. *Journal of Food Science*, **35**, 343–5.

Learmouth, E. M. & Wood, J. C. (1962). Some aspects of soya in food technology. In: *Proceedings of the First International Congress of Food Science Technology*, vol. IV, *Manufacture and distribution of foods* (ed. J. M. Leitch), pp. 785–97. Gordon & Breach Scientific Publishers, New York, 1969.

Liener, I. E. (1972). Nutritional value of food protein products. In: *Soybeans: chemistry & technology*, vol. 1, *Proteins* (ed. A. K. Smith & S. J. Circle), pp. 203–77. Avi Publishing Co., Westport, Connecticut.

Lloyd, J. W. (1940). Range of adaptation of certain varieties of vegetable type soybeans. University of Illinois Agriculture Experimental Station Bulletin No. 471.

Lloyd, J. W. & Burlison, W. L. (1939). Eighteen varieties of edible soybeans. University of Illinois Agriculture Experimental Station Bulletin No. 453.

Lo, W. Y., Steinkraus, K. H., Hand, D. B., Wilkens, W. F. & Hackler, L. R. (1968). Yields of extracted solids in soy milk as affected by temperature of water of various pre-treatments of beans. *Food Technology* **22**, 1322–4.

Concentrates by mechanical extraction

McComb, A. H. (1937). Process for treating soybeans. US Patent 2088853 (August 3).

Maga, J. A. (1971). Indigenous and derived flavor constituents in soy products. In: *Soy: the wonder bean*, pp. 47–68. Symposium, Southern California Section AACC., Aug. 19–21. American Association of Cereal Chemists, St Paul, Minnesota. [Also, *Journal of Agricultural and Food Chemistry*, **21**, 864–8 (1973).]

Moulton, R. H. (1938). Soybean nuts. US Patent 2135592 (November 8).

Mustakas, G. C. (1971). Full-fat and defatted soy flours for human nutrition. *Journal of the American Oil Chemists' Society*, **48**, 815–19.

Mustakas, G. C., Albrecht, W. J. & Bookwalter, G. N. (1972). Production of vegetable protein beverage base. US Patent 3639129 (February 1).

Mustakas, G. C., Albrecht, W. J., Bookwalter, G. N., Sohns, V. E. & Griffin, E. L. Jr (1971). New process for low-cost, high-protein beverage base. *Food Technology*, **25**, 534–8, 540.

Nelson, A. I., Wei, L. S. & Steinberg, M. P. (1971). Food products from whole soybeans. *Soybean Digest* **31** (No. 3), 32–4.

Nohe, I. A. (1938). Soybeans nuts. US Patents 2135593, 2135594 (November 8).

Norris, F. A. (1964). Extraction of fats and oils. In: *Bailey's industrial oil and fat products*, 3rd edition (ed. D. Swern). Wiley-Interscience, New York.

Noyes, R. (1969). *Protein food supplements.* Food Processing Review No. 3. Noyes Data Corp., Park Ridge, NJ.

Orr, E. (1972). *The use of protein-rich foods for the relief of malnutrition in developing countries: an analysis of experience.* Report G73. Tropical Products Institute, London.

Orr, E. & Adair, D. (1967). *The Production of protein foods and concentrates from oilseeds.* Report G31. Tropical Products Institute, London.

PAG (1973). PAG Guideline No. 13 for the preparation of milk substitutes of vegetable origin and toned milk containing vegetable protein. In: PAG Bulletin III No. 1, pp. 14–18. Protein Advisory Group of the United Nations, United Nations, New York.

Pichel, M. J. & Weiss, T. J. (1967). Process for preparing nut butter from soybeans. US Patent 3346390 (October 10).

Piper, C. V. & Morse, W. J. (1923). *The soybean.* McGraw-Hill, New York.

PISFM (1972). *Proceedings of the International Symposium on the Conversion and Manufacture of Foodstuffs by Microorganisms.* (The Sixth International Symposium of the Union of Food Science and Technology, December 5–9, 1971) Saikon Publishing Co., Tokyo.

Rackis, J. J. (1972). Biologically active components. In: *Soybeans: chemistry & technology*, vol. 1, *Proteins* (ed. A. K. Smith & S. J. Circle), pp. 158–202. Avi Publishing Co., Westport, Connecticut.

Rhee, K. C., Cater, C. M. & Mattil, K. F. (1973). Aqueous process for pilot plant scale production of peanut protein concentrate. *Journal of Food Science*, **38**, 126–8.

Robinson, W. B., Bourne, M. C. & Steinkraus, K. H. (1971). Development of soy-based foods of high nutritive value for use in the Philippines. Agency for International Development, Publication PB-213,758. Available from NTIS, US Dept. Commerce, Springfield, Virginia.

Rockland, L. B. (1972). Quick-cooking soybean products. US Patent 3635728 (January 18).

Schlosser, G. C. & Dawson, E. H. (1969). *Cottonseed flour, peanut flour and soybean flour: formulas and procedures for family and institutional use in developing countries.* (USDA-ARS 61–7.) US Dept. of Agriculture, Washington, DC.

Schroder, D. J. & Jackson, H. (1971). Preparation of soybean cheese using lactic starter organisms. 3. Effects of mold ripening and increasing concentrations of skim milk solids. *Journal of Food Science*, **36**, 22–4.

Schroder, D. J. & Jackson, H. (1972). Preparation and evaluation of soybean curd with reduced beany flavor. *Journal of Food Science*, **37**, 450–1.

Scott, W. O. & Aldrich, S. R. (1970). *Modern soybean production.* The Farm Quarterly, Cincinnati, Ohio.

Shemer, M., Wei, L. S. & Perkins, E. G. (1973). Nutritional and chemical studies of three processed soybean foods. *Journal of Food Science*, **38**, 112–15.

Shih, V. E. (1970). Soybean milk. *Journal of the American Dietetic Association*, **57**, 520–2.

Smith, A. K. (1945). Debittering soybeans. A list of patents for removing the bitter taste of soybeans. *Soybean Digest*, **5** (No. 7), 25–6, 28.

Smith, A. K. (1949). *Oriental methods of using soybeans as food with special attention to fermented products.* (ISDA-AIC 234 reprinted as USDA-ARS 71–17 in 1969.) US Dept. of Agriculture, Washington, DC.

Smith, A. K. (1958). *Use of United States soybeans in Japan.* (ISDA-ARS 71–12.) US Dept. of Agriculture, Washington, DC.

Smith, A. K. & Circle, S. J. (1972*a*). Chemical composition of the seed. In: *Soybeans: chemistry & technology*, vol. 1, *Proteins* (ed. A. K. Smith & S. J. Circle), pp. 61–92. Avi Publishing Co., Westport, Connecticut.

Smith, A. K. & Circle, S. J. (1972*b*). Protein products as food ingredients. In: *Soybeans: chemistry & technology*, vol. 1, *Proteins* (ed. A. K. Smith S. J. Circle), pp. 339–88. Avi Publishing Co., Westport, Connecticut.

Smith, A. K., Watanabe, T. & Nash, A. M. (1960). Tofu from Japanese and United States soybeans. *Food Technology*, **14**, 332–6.

Smith, J. M. & Van Duyne, F. O. (1951). Other soybean products. In: *Soybeans and soybean products* vol. 2 (ed. K. S. Markley), pp. 1055–78. Wiley-Interscience, New York.

Sugimoto, H. & Van Buren, J. P. (1970). Removal of oligosaccharides from soy milk by an enzyme from *Aspergillus satoi. Journal of Food Science*, **35**, 655–60.

USDA (1974). *World agricultural production and trade statistical report*, p. 5. FAS (March). US Dept. of Agriculture, Washington, DC.

Van den Berg, C. (1972). Plant protein concentrates. *Deut. Ges. Chem. Apparatew.* (*DECHEMA*), **70**, 27–54.

Van Duyne, F. O. (1950). Recipes for using soy flour, grits, flakes, and soybean oil. University of Illinois Department of Agriculture Circular No. 664.

Watanabe, T. (1969). Industrial production of soybean foods in Japan. In: *UN industrial development organisation meeting on soya bean processing and use*, Peoria, Illinois, Nov. 17–21. United Nations, New York.

Weiss, M. C., Wilsie, C. P., Lowe, B. & Nelson, P. M. (1942). Vegetable soybeans. Iowa State University Agricultural Experimental Station Bulletin P39.

Whiteman, E. F. & Keyt, E. K. (1938). *Soybeans for the table.* USDA Leaflet 166. US Dept. of Agriculture, Washington, DC.

Wilkins, W. F. & Hackler, L. R. (1969). Effect of processing conditions on composition of soy milk. *Cereal Chemistry*, **46**, 381–7.

Williams, L. F. (1950). Structure and genetic characteristics of the soybean. *Soybeans and soybean products*, vol. 1 (ed. K. S. Markley), pp. 111–34. Wiley-Interscience, New York.

Woodruff, S. & Klaas, H. (1938). A study of soybean varieties with reference to their use as food. University of Illinois Agricultural Experimental Station Bulletin No. 443.

9. Rapeseed and other crucifers

R. OHLSON & R. SEPP

Production of rapeseed

Today rapeseed ranks fifth among the oilseeds of the world. Cultivation of the plant for oilseed production is almost entirely confined to the temperate and warm temperate zones. Rape thrives best in rich soil in a cool, moist climate. During the 1960s the production of rapeseed increased from 3.5 to 6.6 megatonnes (Mt) and in 1973 it was up to 7.2 Mt. The only oilseeds to show greater rises were soya and sunflower. The production in different countries is shown in Table 9.1. It seems likely that world rapeseed production will continue to grow; world production in 1985 may reach 12 Mt. The main reason for this is the probability that antinutritional components can be eliminated. Mustard seed is produced in different countries up to some hundred thousand tonnes. The quantities of crambe produced are negligible.

Botany of rapeseed

Rapeseed from Europe generally consists of winter rape (*Brassica napus*) while rapeseed from Canada is usually summer turnip rape (*B. campestris*). Rapeseed from the Far East may be either summer turnip rape, brown mustard (*B. juncea*) or a mixture of these two species (Table 9.2). Most examination keys use the inflorescence to distinguish between the species. *B. napus* has the buds above the just opened flowers, whereas in *B. campestris* the buds are at a lower level. The most reliable character used for distinguishing the species in the generative phase is the shape of the upper leaves. The seeds of *B. napus* are generally larger than those of *B. campestris* and *B. juncea*. Seeds vary from light yellow to brown and black.

Unlike many other important seed plants, the *Brassica* species have very little endosperm, and when the seed coat, responsible for 12–20 % of the seed weight, is removed, the interior contains the embryo. Most of the embryo consists of the cotyledons which contain about 50 % oil and protein-rich grains similar to those in the aleurone cells lying just under the seed coat. Each cell is surrounded by a thin cell wall, mainly composed of cellulose. Lipid droplets are scattered through the cytoplasm. The aleurone grains measure 2–10 μm in diameter.

Table 9.1. *Rapeseed production in different countries in tonnes × 10^3*

	1971	1972	1973
India	1410	1800	—
Canada	2121	1279	1817
China	1000	925	
France	640	703	625
Poland	623	450	750
Sweden	249	327	290
Pakistan	296	285	—
West Germany	225	245	222
East Germany	193	220	200
World total	7410	7044	*c.* 7200

Chemical constituents of rapeseed

The moisture content of 'naturally dry' rapeseed is 6 to 8 %. The oil content of seeds varies. Typical figures for winter rape are 42 to 50 %, for summer rape 37 to 47 %, for winter turnip rape 40 to 48 % and for summer turnip rape 36 to 46 % in the dry matter. Typical fatty acid compositions are given in Table 9.3.

By plant breeding it has been possible to develop cultivars with very low levels of the long-chain fatty acids. Therefore it is likely that within a few years commercial oils will no longer have the traditional composition. The amounts of the predominant glucosinolates in different cultivars of rape are shown in Table 9.4. The finding of a cultivar of summer rape, 'Bronowski', with a markedly reduced level of glucosinolates has a great influence on the development of new varieties. The myrosinase (thioglucoside glucohydrolase) in rapeseed catalyses the hydrolysis of the glucosinolates to give glucose, sulphate, isothiocyanates (ITC), and (L)-5-vinyl-2-oxazolidinethione (VOT). The residue after solvent extraction contains about 40 % protein. Its amino acid composition is given in Table 9.5, and the component proteins in Table 9.6.

Manufacture of rapeseed oil and meal

Rapeseed is not dehulled in commercial plants; the small size and the high oil content of the seeds make it difficult to dehull efficiently. After cleaning, it is crushed by roller mills and heated to complete the breakdown of cells, coagulate the proteins, reduce the affinity of the oil for

Table 9.2. *Definition of rapeseed*

Botanical (Latin) name	Correct English	Synonyms	Canadian	French	German
Brassica napus					
subsp. *oleifera*		Oil rape, rapeseed	Argentine	Colza d'hiver	Winterraps
forma biennis	Winter rape	Swede rape,		Colza de printemps	Sommerraps
forma annua	Summer rape	Oilseed rape		Colza d'été	
Brassica campestris					
subsp. *oleifera*	Winter turnip rate	Rapeseed,	Polish	Navette d'hiver	Winterrübsen
forma biennis	Summer turnip rape	Oil turnip		Navette de printemps	Sommerrübsen
forma annua				Navette d'été	
forma annua var. *chinensis*	Summer turnip rape	Chinese mustard		Moutarde chinoise Pak-choi	Chinasenf
var. *pekinensis*	Summer turnip rape	Celery cabbage Chinese kale		Chou chinois Pet-sai	Chinakohl
var. *dichotoma*	Summer turnip rape	Toria		Toria	Toria
var. *trilocularis*	Summer turnip rape	Sarson		Sarson	Sarson
Brassica juncea	Brown mustard	Leaf mustard Indian mustard	Oriental mustard	Moutarde brune	Brauner Senf Sarepta Senf

From Appelquist & Ohlson (1972).

Table 9.3. *Typical ranges of variation in content of common fatty acids in the oils of some cultivars or breeding lines of some cruciferous seeds*

Species and types	Palmitic	Oleic	Linoleic	Linolenic	Eicosenoic	Erucic
	Ranges in percentage content of					
Brassica campestris						
Winter turnip rape	2–3	14–16	13–17	8–12	8–10	42–46
Summer turnip rape, classical cultivars	2–3	17–34	14–18	9–11	10–12	24–40
Summer turnip rape, low-erucic acid lines†	4–7	48–55	27–31	10–14	0–1	0
Sarson and Toria	2–3	9–16	11–16	6–9	3–8	46–61
Brassica juncea	2–4	7–22	12–24	10–15	6–14	18–49
Brassica napus						
Winter rape, classical cultivars	3–4	8–14	11–15	6–11	6–10	45–54
Winter rape, low-erucic acid cultivars or lines	4–5	40–48	15–25	10–15	3–19	3–11
Summer rape, classical cultivars‡	3–4	12–23	12–16	5–10	9–14	41–47
Summer rape, low-erucic acid cultivars or lines†	5	52–55	24–31	10–13	0–2	0–1
Brassica tournefortii	2–4	6–12	11–16	10–16	6–8	46–52
Sinapis alba	2–3	16–28	7–10	9–12	6–11	33–51

† Results from only a few samples available. The range of variation in fatty acid compositions is expected to increase as new cultivars are being released.

‡ Except for the Polish cultivar Bronowski, which is not grown to any significant extent. This cultivar has *c.* 10 % erucic acid.

From Appelquist & Ohlson (1972).

solid surfaces, make the phosphatides insoluble, increase the fluidity of the oil, destroy moulds and bacteria, inactivate enzymes and dry the seed. Inactivation of myrosinase is the most important of these objectives.

The cooking is generally carried out in a series of closed cylindrical steel kettles stacked on top of each other. Each kettle is equipped with a sweep-type stirrer. The kettles are also jacketed for steam heating. The length of time in the cooker is usually 30 to 60 min, and the temperature 75 to 120 °C.

Screw (expeller) presses are used either for straight pressing, or more commonly pre-pressing for subsequent solvent extraction. After pre-pressing, the crushed cakes are flaked before solvent extraction. This step is very important because the seed particles must be thin enough to be extracted readily, and yet large enough to form a mass through which the solvent will freely flow. Extraction of rapeseed with solvents is the most efficient method of obtaining the oil. Generally, pre-pressing

Table 9.4. *Content of glucosinolates in seed meals of some European cultivars of rape and turnip rape, grown at various localities*

Species and cultivars	Origin	Glucosinolate content (% in dry matter) Gluconapin	Progoitrin	Gluconapin + progoitrin
Brassica napus, winter type				
Matador	Mean of 3 samples, grown in Sweden	1.71	4.69	6.40
Heimer	Mean of 7 samples, grown in Sweden	1.71	4.51	6.24
Victor	Mean of 11 samples, grown in Sweden	1.79	4.31	6.11
Brassica napus, summer type				
Regina II	Mean of 8 samples, grown in Sweden	1.27	3.04	4.31
Rigo	Mean of 3 samples, grown in Sweden	1.31	3.72	5.03
Bronowski, I.L. 1997	Unknown (received from Poland)	0.15	0.79	0.94
Brassica campestris, winter type				
Duro	Mean of 3 samples, grown in Sweden	3.10	0.22	3.32
Rapido II	Mean of 3 samples, grown in Sweden	2.95	0.50	3.44
Brassica campestris, summer type				
Bele	Mean of 2 samples, grown in Sweden	1.78	1.26	3.30

From Applequist & Ohlson (1972).

is used to give a press-cake containing 12 to 20 % oil. The extraction gives a residual oil content of about 0.5 %. The most widely used solvent for extraction is commercial *n*-hexane. The solvent to meal solids ratio in modern extractors is about 1.1 to 1.3. After completed solvent extraction the hexane is removed by indirect heating and by direct steam injection. Besides leaving a meal with not more than 0.01 % hexane, this step should give the meal an appropriate heat treatment, so that the feeding value of the meal will remain high.

Table 9.5. *Amino acid composition in rapeseed protein concentrate*

Amino acid	g/16 g N	Amino acid	g/16 g N
Isoleucine	4.4	Histidine	1.5
Leucine	7.9	Arginine	7.2
Lysine	6.7	Aspartic acid	8.0
Phenylalanine	3.8	Glutamic acid	18.8
Tyrosine	3.2	Serine	5.0
Cystine	2.0	Proline	6.1
Methionine	2.2	Hydroxyproline	0.7
Threonine	4.7	Glycine	5.4
Valine	5.6	Alanine	4.9
Tryptophan	1.6		

Table 9.6. *Rapeseed proteins*

Molecular weight	Number of proteins	Isoelectric point	% of total proteins
320000	5–7	4–7	15–20
150000†	10–20	5–8	50–60
75000	3–5	5–8	3–5
13000	5	11	15–20

† Containing the myrosinases (low percentage).

Table 9.7. *Composition of rapeseed meal and soya meal*

	Dry matter %	Chemical composition, % of dry matter: Protein (N×6.25)	Fat	N-free extract	Crude fibre	Ash	Ca	P
Rapeseed meal								
B. napus	89	43.1	2.3	36.9	10.7	7.0	0.8	1.2
B. campestris	89	41.1	2.0	37.8	11.6	7.5	0.6	1.3
Sinapis alba	89	45.5	1.2	36.2	9.8	7.3	0.8	1.2
Soya meal	87	50.4	0.5	35.4	6.9	6.8	0.3	0.7

From Appelquist & Ohlson (1972).

Nutritional value and use of rapeseed meal

Composition

Some data of the chemical composition of rapeseed and soybean meals are given in Table 9.7. Although the protein content of rapeseed meal is less than that in soya meal, it is the most concentrated vegetable protein feed produced in the cool temperate zones. The protein content is increased by nitrogen fertilisers or abundant rain. The amino acid

Table 9.8. *Amino acid composition of rapeseed meals, produced by different methods and in different periods, as percentage of protein*

	1956–61		1965–7	
Amino acid	Expeller type (Mean of 5 samples)	Pre-press–Solvent type (Mean of 15 samples)	Pre-press–Solvent type (Mean of 8 samples)	Solvent type (Mean of 7 samples)
Arginine	5.09	5.48	5.59	5.65
Histidine	2.40	2.66	2.66	2.63
Lysine	4.39	5.31	5.94	5.85
Tyrosine	2.16	2.10	2.30	2.28
Tryptophan	0.94	1.21	1.23	1.29
Phenylalanine	3.74	3.78	3.83	3.89
Cystine	0.73	0.80	1.19	1.22
Methionine	1.88	1.91	1.87	1.83
Serine	4.03	4.19	4.40	4.37
Threonine	4.08	4.19	4.41	4.35
Leucine	6.45	6.67	6.95	7.04
Isoleucine	3.71	3.63	3.81	3.87
Valine	4.76	4.82	4.94	4.98
Glutamic acid	16.16	16.84	18.71	18.99
Aspartic acid	6.58	6.72	7.02	7.04
Glycine	4.68	4.78	5.08	5.09
Alanine	4.21	4.29	4.44	4.46
Proline	5.71	6.13	6.13	6.32

From Appelquist & Ohlson (1972).

composition of the commercial meal is largely influenced by the way in which the meal is produced from the seed. High temperatures and pressures result in extensive destruction of lysine. In modern processing, a combination of pre-pressing and solvent extraction preserves the basic amino acids, as shown in Table 9.8. The lysine content of rapeseed is about the same as in soya meal, whereas the cystine and methionine contents are higher.

The main obstacle to the use of rapeseed meal as feed seems to be its content of glucosinolates. These compounds are easily soluble in water, ethanol, methanol, and acetone. The split products – ITC and VOT – are soluble in ethyl ether and in chloroform. Steam-volatile ITCs are lipophilic and soluble, for instance in hexane, while the non-volatile ITCs are less lipophilic, and the VOTs non-volatile and relatively hydrophilic.

Antinutritional effects

The harmful effect most often observed from regular feeding of rapeseed meal is thyroid enlargement. This effect cannot be entirely counteracted by feeding iodine. Autolysed meals containing nitriles also cause decreased weight gain and enlargement and microscopic lesions in the liver and kidneys. Many methods are used or have been suggested to counteract this toxicity, such as enzyme inactivation or removal or destruction of glucosinolates or cleavage products. Examples of these methods are: autolysis and distillation; autoclaving and steam stripping; treatment with ammonia and heat; treatment with ferrous salts; and extraction of toxic factors. The best method, however, seems to be the removal of glucosinolates by plant breeding.

Feeding value

Since the cleavage products of glucosinolates do not exhibit toxic effects in ruminants, rapeseed meal is mainly used for them. The digestibility of rapeseed meal and the hull fraction is good. For non-ruminant animals such as swine, the amino acid composition of the protein feed is more important than for ruminants. Rapeseed meal is in this respect very favourable, but swine seem to be relatively sensitive to the cleavage products of glucosinolates.

It is known from several reports that feeding more than 5 % of rapeseed meal to poultry causes thyroid enlargement. It is possible that seed meal of *B. napus* should not be fed to swine or poultry at as high a level as meal of *B. campestris*. The new cultivars containing a reduced amount of glucosinolates will, when grown on a large scale, increase the use of rapeseed meal.

Processes for producing edible rapeseed protein concentrate (RPC)

By analogy with soybean technology the rapeseed protein concentrate may be defined as 'the product prepared from high quality, sound, clean dehulled seeds by removing most of the oil and water-soluble non-protein components' (Smith, 1972). In rape, mustard and crambe technology, special attention is given, therefore, to the complete removal of the antinutritional components, i.e., glucosinolates. Table 9.9 presents an inventory of different approaches possible; they have not all been tested; neither does each approach lead solely and directly to a market-

Table 9.9. *Review of process steps for detoxification and concentration of seed protein*

Method	Intact seed	Ground seed	Seed milk	Defatted meal
Dry methods		Dehulling		Dehulling by air classification
Heat treatment	Myrosinase inactivation (Eapen *et al.*, 1968) (Dahlên & Goude, 1973)	Steam stripping (Mustakas *et al.*, 1965)*	Yuba imitation	Steam stripping (Goering, 1961)* Toasting (Rutkowski & Kozlowska, 1967)
Extraction and other solid–liquid separation	Diffusion extraction (van Etten *et al.*, 1969)† (Sosulski *et al.*, 1972)	FRI process (Tape *et al.*, 1970) Dehulling by various wet methods Defatting by various commercial processes	Defatting by skimming Ultrafiltration Curd precipitation	Soy protein concentrate analogue (Ballester, 1970; Lo & Hill, 1971) Alcohol extraction (Frölich, 1953; Nehring, *et al.*, 1967)‡
Biochemical treatment	Germination Dehulling by cellulase activity Tempeh imitation	Warm fermentation (Mustakas, 1963)*	Cheese imitation Sufu imitation	Fermentation (Staron, 1970) (Poznanski *et al.*, 1973)

* Mustard. † Mustard and crambe. ‡ Crambe and rape.

able protein concentrate. Commercial approval of one method or another will depend on many factors. Generally the removal of hulls will improve the colour, increase the protein content and decrease the fibre content of the proteinaceous residue.

If the dehulling step precedes pressing and extraction of the oil, there is an almost 100 % hull removal. However, several complications are inevitable. As a typical hull fraction contains 7 % of total seed lipids, it should be oil-extracted and this would necessitate an additional process step. Furthermore, absence of hulls in the press-cake may demand a slightly different process or equipment for oil pressing and extraction. On the other hand, if the dehulling is carried out by, e.g., air classification of the defatted meal (in order to avoid losses in oil yield), it will be impossible to achieve total hull removal and the nutritional quality of the RPC may be considerably lower. Thus, conditions giving maximum oil yield increase the fibre in the product.

Concentrates by mechanical extraction

The decision about the optimum place for the dehulling step in the manufacturing process will consequently depend on economic factors. In a conventional oil mill, the production is optimised according to oil yield and quality. Consequently, meal production runs at a suboptimal level. Some factors influencing the choice of the best method are as follows.

Temperature

Heat inactivates enzymes such as myrosinase and facilitates the extraction of non-protein components. Heat also reduces the solubility of proteins and, unfortunately, decreases the oil quality (by oxidation, polymerisation, discolouring) and raises the operating costs.

Seed disintegration

The disadvantages of extracting intact seed (Kozlowska & Sosulski, 1972) are the long extraction time, large volumes of water needed, and increased heat consumption. These factors may lead to the opinion that the process is economically hazardous. Advantages are limited material losses and simplicity in drying the intact seed. The main disadvantages in the extraction of ground seed are the great material losses (decreased oil and protein yields), and the expense of waste treatment. Such a process cannot be justified economically.

Influence of chemicals

Adding alkaline water before oil recovery may have adverse effects. Some sulphur from glucosinolates seems to diffuse into the oil phase (Kozlowska & Sosulski, 1972) and part of the protein may be dissolved in the aqueous phase. Ethanol may be used as the solvent to extract non-protein components. One advantage of this is the inactivation of myrosinase, another being a decreased sulphur content in the oil. A great disadvantage is the cost of ethanol.

A new process

An ideal solution may be the process now developed in Sweden, which comprises four main steps: dehulling, myrosinase inactivation by heat treatment, glucosinolate removal by water leaching, and oil extraction. Effluents are recirculated to diminish losses (Ohlson, 1973).

Properties of rapeseed protein concentrates

The RPC has a composition (Table 9.10) well suited to a product for human consumption, with a protein content of approximately 60 % and a moisture content of 8 %. Rapeseed protein has an amino acid composition (Table 9.5) better than that of soybeans, and one which compares well with many animal proteins. Unlike soya protein, the important amino acids methionine and cystine are present in adequate quantities. The good amino acid composition is also evident in results of rat feeding studies where the following data were obtained:

Protein efficiency ratio (PER)	2.8–3.8	(2.5–3.0 when adjusted to casein with a PER value of 2.5)
Net protein utilisation (NPU)	78	
Biological value (BV)	92	
Digestibility	86	

Table 9.10. *Typical analysis of RPC (rapeseed protein concentrate) from* B. napus, *winter type*

Component	Content in dry matter (%)
Protein (N × 6.25)	65
(Protein (N × 5.5)	57
Fat	1
Carbohydrates (excl. fibres)	28
Crude fibres	7
Total ash	7
Glucosinolates	< 0.03

The water-absorption capacity of RPC is 500 to 700 %, as compared to 400 % for texturised soybean flour and the water-holding capacity 400 % as compared to 200 to 300 %. RPC can also bind fat, which is useful for regulating the fat content in foods. RPC could be used at 2 to 3 times the concentration of soybean proteins with no detectable off-flavour. The solubility of the protein in RPC is small and comparable to that of heat-treated soybean products. The nitrogen solubility index of RPC is about 15 in pure water and about 25 in 0.2 M NaCl. The emulsifying and foaming properties are the same as for heat-treated soybean products.

The RPC is intended for use as an addition to the ordinary diet. It could be used in a great many foods such as minced meat products, sausages, canned meat products, bread, biscuits, etc. Preliminary studies for 3 months on rats have shown that the material is safe and no toxic effects from the small amount of residual glucosinolates could be detected. Extensive studies of RPC with two animals and for periods up to 2 years are in progress. After this, the test results will be presented to national and international authorities for approval.

Processes of producing rapeseed protein isolate (RPI)

RPI is not yet commercially manufactured. It is a product which contains the protein fraction and a minimum amount of the non-protein components of the seeds. The products should be prepared from high quality raw material and should have a protein content exceeding some arbitrarily chosen lower limit, e.g. 90 %. It is necessary to use processes resembling those adopted in the preparation of protein isolates from soybeans in which the protein is dissolved and recovered from solution by precipitation.

Protein yield in an isolate process may vary greatly depending on factors such as the extent of heat treatments during meal processing and the necessity of protein-washing steps after precipitation. In the case of a commercial meal, (Owen, 1971) 18 % of original seed nitrogen was recovered (giving an isolate with a protein content of 84 %). However, under laboratory conditions, the reported protein yield (Sosulski *et al.*, 1972) may be almost 50 % (RPI with 86 % protein). This was confirmed by our own laboratory where yields of 29 to 49 % were achieved of RPI containing more than 90 % protein (Kroon *et al.*, 1970). An important problem is to obtain a product that is entirely free from glucosinolates and products from glucosinolate decomposition.

A Swedish team working on the preparation of RPI has reported that in order to get a product suitable for spinning one also has to remove all phytic acid (Törnell *et al.*, 1972). The demand for total removal of these compounds will increase the need for efficient washing of the protein curd. The use of an isolate may thus be limited by increased production costs caused by reduced isolate yields, relatively high washing cost and an increased amount of waste.

An alternative would be to start with a purified rapeseed meal or the RPC. Kodagoda *et al.* (1973) proposed a three stage procedure starting from a protein concentrate. Three distinct protein fractions with different properties are obtained. However, 40 % of the protein in RPC was not recovered.

Treatment of rapeseed protein plant wastes

Wastes may be divided into solid wastes and waste effluents. An example of solid wastes are hulls, which in the case of soybean are considered as ingredients for animal feeds. Waste effluents are more serious. For every tonne of isolated soyprotein produced, in terms of BOD (biological oxygen demand), the waste (water) would be equivalent to that produced by a city of about 10000 people (Johnson, 1969). The waste load per tonne of finished soybean protein concentrate will be about half as much. The waste effluent from a RPC or RPI plant may in a first stage be purified by centrifugation, sedimentation or filtration in order to remove all fine fibres and colloidal protein particles. These can be dried for example, together with hulls or other solid wastes. The filtrate, depending on the extraction procedure used, will contain different proportions of carbohydrates, minerals, lipids, proteins (with peptides and amino acids) and glucosinolates and has to be purified in a second stage.

In a recent study (Kozlowska *et al.*, 1972), extraction procedures on intact and ground seeds were compared with leaching of defatted seeds. It is obvious that the wasted dry matter will be most abundant in the treatment of ground seeds. The effluent will then contain 20–30 % oil in the dry matter and will have to be treated like wastes from slaughter houses and oil mills. In the case of diffusion extraction (Table 9.9) the oil portion is low. The waste water will be mostly a carbohydrate-containing stream, similar to brewery wastes. In the defatted meal process, treatments will probably be similar to those used on cheese whey or potato fruit water.

Common for all effluents from a rape, crambe or mustard seed protein plant will be the content of glucosinolates and their hydrolysis products. Micro-organisms may be used for detoxification of glucosinolate-containing solutions. Rutkowski *et al.* (1972) have used an extract of rapeseed meal containing most of glucosinolates for the microbiological synthesis of vitamin B_{12}.

References

Appelquist, L. Å. & Ohlson, R. (1972). *Rapeseed.* Elsevier Publishing Co., Amsterdam.

Ballester, D. (1970). *Journal of the Science of Food and Agriculture*, **21**, 143.

Bhatty, R. S. & Sosulski, F. W. (1972). *Journal of the American Oil Chemists' Society*, **49**, 346.

Concentrates by mechanical extraction

Dahlén, J. & Goude, A. (1973). Swedish Patent No. 357658.

Eapen, K. E., Tape, N. W. & Sims, R. P. A. (1968). *Journal of the American Oil Chemists' Society*, **45**, 194.

Frölich, A. (1953). *Kungliga Lantbrukshögskolan Ann.*, **19**, 205.

Goering, K. J. (1961). (For Oilseed Produces Inc.) US Patent No. 2987399.

Johnson, D. W. (1969). At: UNIDO Expert Group Meeting, Peoria, 17–21 Nov.

Kodagoda, L. P., Yeung, C. Y., Nakai, S. & Powrie, W. D. (1973). *Canadian Institute of Food Science and Technology Journal*, **6**, 135.

Kozlowska, H. & Sosulski, F. W. (1972). *Canadian Institute of Food Science and Technology Journal*, **5**, 149.

Kroon, S. E., Sepp, R. & Teär, J. (1970). Report on protein food promotion. Kasetstart University, Bangkok, 22 Nov.–1 Dec.

Lo, M. T. & Hill, D. C. (1971). *Journal of the Science of Food and Agriculture*, **22**, 128.

Mustakas, G. C. (1963). *Biotechnology and Bioengineering*, **5**, 27.

Mustakas, G. C., Kirk, L. D., Sohns, V. E. & Griffin, E. L., Jr (1965). *Journal of the American Oil Chemists' Society*, **42**, 33.

Nehring, K., Bock, H.-D. & Wünsche, J. (1967). *Sitzungberichte, DDR Deutsche Akademie der Landwirtschaftswissenschaft, Berlin*, **16**, 103.

Ohlson, R. (1973). PAG Bulletin No. 3. Protein Advisory Group of the United Nations, Rome.

Owen, D. F. (1971). *Cereal Chemistry*, **48**, 91–6.

Poznanski, S., Bednarski, W. & Jakubowski, J. (1973). *Lait*, **53**, 169.

Rutkowski, A. *et al.* (1972). *Canadian Institute of Food Science and Technology Journal*, **5**, 67.

Rutkowski, A. & Kozlowska, H. (1967). *Sruta Rzepakowa*. Warsaw. (In Polish.)

Smith, A. K. & Circle, S. J. (1972). *Soybeans: chemistry and technology*, vol. 1, pp. 440–3. AVI Publishing Co, Westport, Connecticut.

Sosulski, F. W. & Bakal A. (1969). *Canadian Institute of Food Science and Technology Journal*, **2**, 28–32.

Sosulski, F. W. & Kozlowska, H. (1972). At: American Oil Chemists Society Meeting, Ottawa, 24–8 Sept.

Sosulski, F. W., Soliman, F. S. & Bhatty, R. S. (1972). *Canadian Institute of Food Science and Technology Journal*, **5**, 101.

Staron, T. J. (1970). In: Proceedings of the International Rapeseed Conference, Sainte Adèle, Canada, p. 321.

Tape, N. W., Sabry, Z. I. & Eapen, K. E. (1970). *Canadian Institute of Food Science and Technology Journal*, **3**, 78.

Törnell, B., Gillberg, L. & Larsson, L. (1972). STU (Swedish National Board for Technological Development) Dokumentation Rapport-referart No. 7, p.18.

van Etten, C. H., Daxenbichler, M. E. & Wolff, I. A. (1969). At: AOCS Meeting, New York, 7–12 Sept.

10. Sunflower, safflower, sesame and castor protein*

ANTOINETTE A. BETSCHART, C. K. LYON &
G. O. KOHLER

Even though the total production of these four oilseeds does not equal that of soy, they are important sources of protein in the areas where they are grown. Production of sunflower has been expanding in recent years so that it is now a major source of edible vegetable oil. The value of these oilseed proteins can be greatly increased by upgrading from fertilizer to feed or from feed to food uses. Because of its toxic components, castor meal is not a food protein, but after detoxification it has great potential for use in animal feeds.

Sunflower

The sunflower (*Helianthus annuus*) is believed to have originated in either North America or Peru. The Spaniards introduced it to Europe during the sixteenth century, but it was not developed as an oilseed crop until the early nineteenth century (Gandy, 1972). Recently it has become an important oilseed crop in the USSR and S.E. Europe mainly because of the development of high oil, high-yielding varieties which can be mechanically harvested (Glicksman, 1971; Huffman *et al.*, 1973). Sunflowers are produced mainly for their oil which is highly unsaturated and has excellent stability (Kromer, 1967; Gandy, 1972). The bird feed and human food markets are of lesser importance (Robertson, 1972). The defatted meal which remains after solvent extraction has been used primarily as an animal feed (Clandinin, 1958). Protein concentrates or isolates which have potential as human food have been prepared from the meal (Agren & Lieden, 1968; Gheyasuddin *et al.*, 1970*a*; Burns *et al.*, 1972; Sosulski *et al.*, 1972).

The pertinent literature relating to sunflower has been compiled into a useful bibliography (Posey, 1969). Others have reviewed aspects such as production and yield (Kromer, 1967), processing and utilization in feeds and foods (Clandinin, 1958; Burns *et al.*, 1972), nutritional value (Smith, 1968) and future potential (Robertson, 1972).

* Literature reviewed through 1972.

Concentrates by mechanical extraction

Production and Yield

The world production of sunflower increased from 5.4 to 9.7 megatonnes (Mt) between 1960 and 1970 (FAO, 1971; Ohlson, 1972). Sunflower was fourth in the world as a source of edible vegetable oil in 1960 but in 1970 it ranked second to soy (Earle *et al.*, 1968; Gandy, 1972). The major increase in production was in the USSR and Eastern Europe which accounts for 80 % of world production (Kromer, 1967; Gandy, 1972).

The average yield throughout the world between 1968 and 1970 was 1130–1280 kg/ha (FAO, 1971). Yields ranging from 1120–2250 kg/ha were reported for some regions in the US (Kromer, 1967; Trotter & Givan, 1971; Robertson, 1972). An indication of the comparative yields for sunflower, soy and flaxseed are given in Table 10.1 (Kromer, 1967).

Composition

By selection the USSR workers have increased the oil content of the achene from 25–30 % to 50–5 % (Panchenko, 1966; Kromer, 1967; Gandy, 1972). In general, high oil varieties contain 17–25 % protein and 34–45 % oil (Clandinin, 1958; Robertson *et al.*, 1971). The kernels consist of approx. 50 % oil, 26–33 % crude protein, 9–12 % crude fiber and 4–6 % total sugars (Pomenta, 1970; Huffman *et al.*, 1973). They are also a good source of B complex vitamins, β-carotene, calcium, iron and phophorus (Huffman *et al.*, 1973). Cancalon (1971) reported that hulls were approx. 50 % cellulose and lignin, 25.7 % reducing sugars, 5.17 % fat and 4 % protein. These data emphasize the importance of decortication if high quality meals are desired.

The protein content of sunflower meal compares favorably with other oilseed meals (Table 10.2). Sunflower meal has a higher content of good quality protein ($\simeq$ 50 %) than most cereals which are used as the major protein source for many countries (Glicksman, 1971; Burns *et al.*, 1972).

Preparation of protein concentrates and isolates

Protein concentrates and isolates have been successfully prepared from sunflower meal or kernels (Agren & Lieden, 1968; Gheyasuddin *et al.*, 1970*a*; O'Connor, 1971; Sosulski *et al.*, 1972). Swedish workers have prepared a flour from solvent-extracted kernels and a heat-coagulated protein concentrate from the aqueous suspension of the flour (Agren &

Table 10.1. *Yield of sunflower, soybean and flaxseed (kg/ha)*

	Oilseed	Oil	Meal
Sunflower	1120	450	390
Soybean	1680	300	1320
Flaxseed	630	225	400

Table 10.2. *Average composition of solvent-extracted oilseed meals*

	Sunflower	Cottonseed	Soybean
Moisture	7.0	9.0	11.0
Ash	7.7	6.5	5.8
Crude fiber	11.0	11.0	6.0
Fat	2.9	1.6	0.9
Protein	46.8	41.6	45.8

Lieden, 1968). The concentrate 52–55 % protein, 2–4 % fiber) was incorporated into a high protein formulation based on tef.

There are two major objectives in the preparation of light-coloured sunflower protein isolates: (1) extraction of the protein with a minimum of denaturation, and (2) removal of phenolic substances which impart a greenish-brown colour to the isolate. Factors which influence the extractability of sunflower protein were studied (Gheyasuddin *et al.*, 1970*b*). At pH 7.0 the protein was more thoroughly extracted by 1.0 M NaCl or 0.75 M $CaCl_2$ than by water (Fig. 10.1). This reflects the presence of larger quantities of globulins (56 %) than albumins (20–2 %) (Sosulski & Bakal, 1969; Burns *et al.*, 1972).

Girault *et al.*, (1970) extracted non-protein nitrogen (N) from solvent extracted seeds with 10 % NaCl buffered at pH 8.6; protein N was then extracted with a borate buffer, pH 10, containing 0.1 % mercaptoethanol and 0.5 % lauryl sodium sulfate. The protein was freed of the green pigments by using Sephadex G-50. O'Connor (1971) prepared a light-colored protein from sunflower meal by extracting the protein in an alkaline solution, passing the extract through an ultrafiltration membrane, and precipitating the protein with acid. Protein is recovered from the retentate which is free of green color-forming precursors. All operations are performed in an inert atmosphere. Gheyasuddin *et al.* (1970*a*) extracted defatted sunflower meal at pH 10.5 in the presence of 0.2 % aqueous sodium sulfite, precipitated the protein at pH 5.0 and extracted the precipitate twice with 50 % aqueous isopropanol. The nearly-

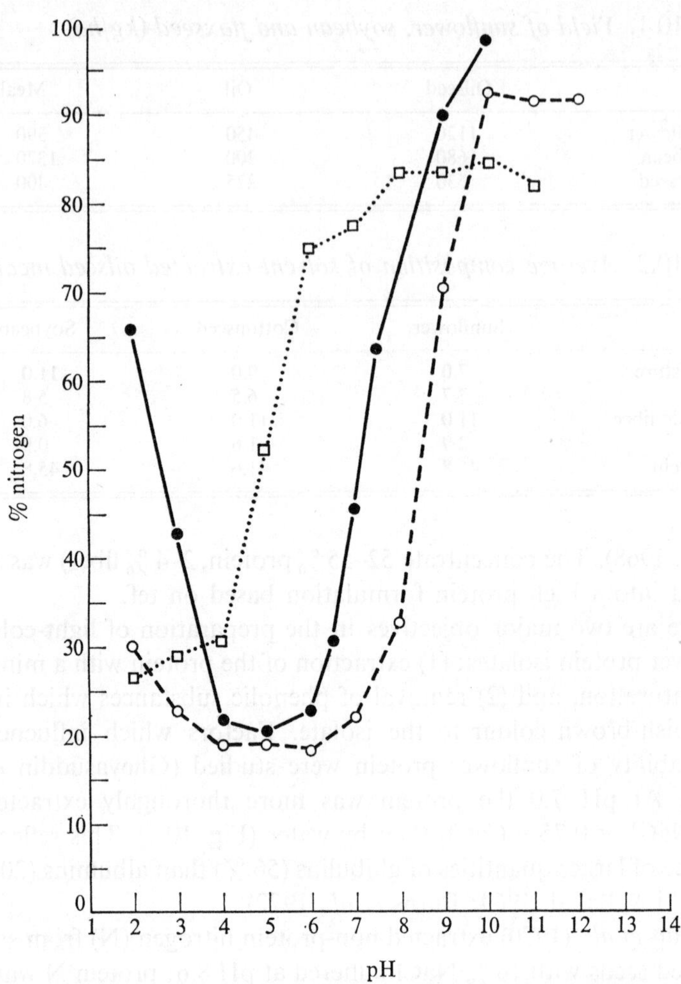

Fig. 10.1. Solubility profiles of sunflower seed meal in water (○--○) and 1.0 M NaCl (□ ··· □) and solubility of isolated sunflower protein after precipitation (●—●).

white isolate (95.9 % protein) still contained constituents which imparted a brown color to the solution above pH 7.5. As reflected in the N solubility profile of the unextracted and isopropanol-extracted isolate (Fig. 10.2), slight denaturation resulted from the solvent extraction. Most recently Sosulski *et al.* (1972) described a method whereby polyphenols and pigments were diffused from intact kernels into an aqueous acid (0.001 N HCl) before extracting the protein. Due to the high temperature involved (80 °C for 6 h) the protein was denatured. The isolate,

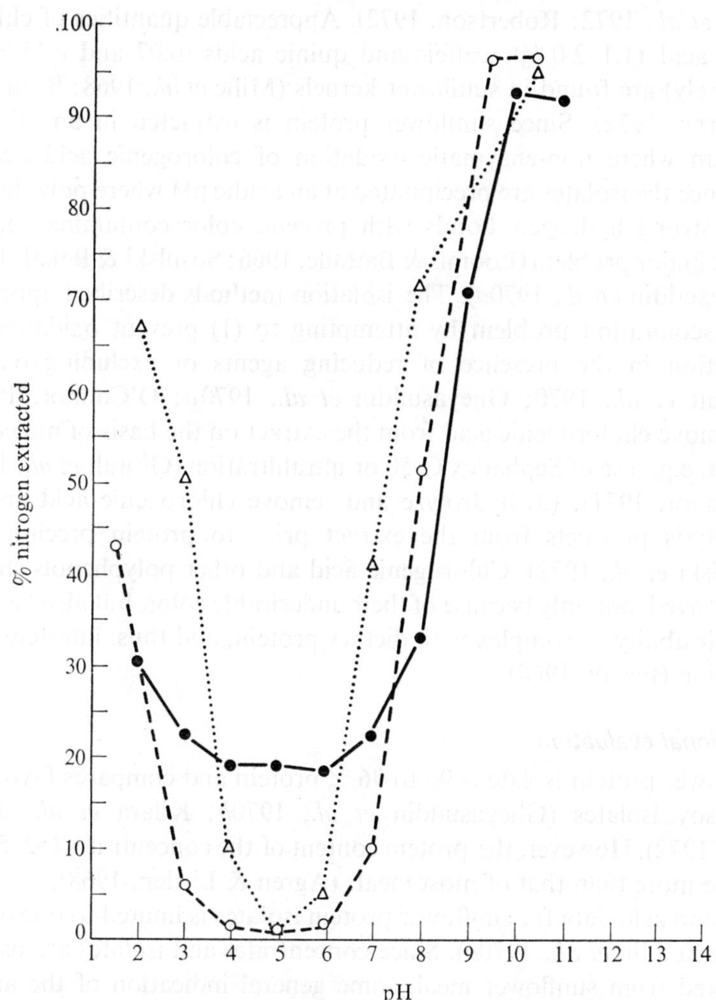

Fig. 10.2. Solubility profiles of sunflower seed meal (●– –●), sunflower protein isolate (△···△), and sunflower protein isolate extracted with 50 % aqueous isopropanol (○– –○).

prepared by isoelectric precipitation was 98.3 % protein and 1.7 % ash. The N solubility profiles of the protein isolates showed that diffusion extraction of the kernels at 60 °C rather than 80 °C resulted in less denaturation, but extraction of chlorogenic acid was less complete (Kilara *et al.*, 1972).

The dark greenish-brown color associated with the sunflower protein isolate is due to the oxidation of chlorogenic acid which coprecipitates with the protein (Sechet-Sirat *et al.*, 1959; Brummett & Burns, 1972;

Cater *et al.*, 1972; Robertson, 1972). Appreciable quantities of chlorogenic acid (1.1–2.0 %), caffeic and quinic acids (0.07 and 0.15 % respectively) are found in sunflower kernels (Milic *et al.*, 1968; Brummett & Burns, 1972). Since sunflower protein is extracted in an alkaline medium where non-enzymatic oxidation of chlorogenic acid occurs, and since the isolates are precipitated at an acidic pH where polyphenols form strong hydrogen bonds with protein, color contamination has been a major problem (Loomis & Battaile, 1966; Sosulski & Bakal, 1969; Gheyasuddin *et al.*, 1970*a*). The isolation methods described approach the discoloration problem by attempting to (1) prevent oxidation by extraction in the presence of reducing agents or excluding oxygen (Girault *et al.*, 1970; Gheyasuddin *et al.*, 1970*a*; O'Connor, 1971); (2) remove cholorogenic acid from the extract on the basis of molecular weight, e.g., use of Sephadex G-50 or ultrafiltration (Girault *et al.*, 1970; O'Connor, 1971); (3) hydrolyze and remove chlorogenic acid and its hydrolysis products from the extract prior to protein precipitation (Sosulski *et al.*, 1972). Chlorogenic acid and other polyphenols should be removed, not only because of their undesirable color, but also because of their ability to complex with dietary protein, and thus, interfere with digestion (Feeny, 1968).

Nutritional evaluation

Sunflower protein isolate is 90 to 96 % protein and compares favorably with soy isolates (Gheyasuddin *et al.*, 1970*b*; Kilara *et al.*, 1972; Wolf, 1972). However, the protein content of the concentrates (52–55 %) is little more than that of most meals (Agren & Lieden, 1968).

Amino acid data for sunflower protein isolates is limited to one report (Gheyasuddin *et al.*, 1970*a*). Since concentrates and isolates are usually prepared from sunflower meal, some general indication of the amino acid composition can be gleaned from data relating to meal. Smith (1968) and Clandinin (1958) have thoroughly reviewed the nutritive value of sunflower kernels and meal. The first limiting amino acids of sunflower protein are lysine and isoleucine for humans and lysine for growing pigs (Earle *et al.*, 1968; Burns *et al.*, 1972). With the exception of lysine the amino acid composition of sunflower compares favorably with soy and cottonseed (Table 10.3). Although sunflower protein isolates contain less cysteic acid and more leucine, phenylalanine and tyrosine than the original meal, their lysine content is the same. Solvent extraction of the isolate with 50 % aqueous isopropanol had no effect on the essential amino acid composition (Gheyasuddin *et al.*, 1970*a*).

Table 10.3. *Amino acid composition of oilseed meals (g/16 g N)*

Amino acid	FAO reference protein*	Sunflower seed (mean and range for 7 varieties)†	Soybean‡	Cottonseed‡
Arginine	—	8.91 (8.4–9.2)	7.6	10.2
Histidine	—	2.47 (2.4–2.6)	2.4	2.7
Isoleucine	4.2	3.97 (3.9–4.1)	5.5	4.1
Leucine	4.8	6.13 (6.0–6.2)	7.7	5.7
Lysine	4.2	3.77 (3.4–4.2)	6.3	4.3
Methionine	2.2	1.91 (1.7–2.1)	1.3	1.2
Phenylalanine	2.8	4.70 (4.6–4.8)	4.9	5.3
Threonine	2.8	3.18 (3.0–3.4)	3.9	3.2
Tryptophan	1.4	1.11 (1.0–1.2)	1.4	1.4
Valine	4.2	4.76 (4.3–5.1)	5.3	4.8

* FAO (1965). † Earle *et al.* (1968). ‡ Rutkowski (1971).

Although sunflower protein is low in lysine, it is superior to most vegetable proteins in digestibility (90 %) and comparable in biological value (60 %) (Clandinin, 1958; Smith, 1968; Gheyasuddin *et al.*, 1970*b*). When sunflower was fed as the sole source of protein at the 10 % level in the diet of rats, it was superior to sesame, comparable to peanut and inferior to soy protein (Table 10.4; Evans & Bandemer, 1967). With adequate lysine and methionine supplementation, the protein nutritive value for sunflower was 85. Agren & Lieden (1968) reported protein efficiency ratios (PER) of 1.30–1.44 (values adjusted to PER 2.50 for casein–methionine reference diet) for the heat-coagulated concentrate. The PER was increased to 1.7 (values adjusted as above) with 0.78 % lysine supplementation. Sunflower protein should either be supplemented with lysine or fed with other high lysine proteins (e.g., soy, milk, fish).

The deleterious effects of high processing temperatures on the nutritive value of sunflower protein have been reviewed (Clandinin, 1958; Smith, 1968). At temperatures of 115–125 °C and above, between 30 and 40 % of the lysine is destroyed (Clandinin, 1958; Bandemer & Evans, 1963; Smith, 1968). These studies emphasize the importance of using milder treatments in the preparation of sunflower meal.

Sunflower meal has been described as being free of toxic substances (Smith, 1968; Gandy, 1972). Other workers, however, have reported the presence of an arginase inhibitor (Reifer & Augustyniak, 1968) and a trypsin inhibitor (Agren & Lieden, 1968). The arginase inhibitor consists mainly of phenolic compounds ($\simeq$ 80 %); chlorogenic acid and

Table 10.4. *Relative nutritive values of seed proteins for rats. Values given as mean ± standard error*

Seeds	Protein nutritive value†
Peanuts (*Arachis hypogea* L.)	
Spanish	44±5
NC-4X	49±8
Virginia early runner	45±9
Alabama early runner	32±13
Safflower (*Carthamus tinctorius* L.)	
P-1	49±9
US-10	63±20
Gila	38±8
Sesame (*Sesamum indicum* L.)	
American K10	15±3
American white	25±13
Brazilian	42±6
Mexican	34±11
Renner 15	30±6
Sunflower (*Helianthus annuus*)	
Arrowhead	45±15
Mennonite	44±10
Greystripe	48±13
Soybeans‡ (*Glycine max.* L.)	
Chippewa	93±14
Harosoy	99±16

† Protein nutritive value = $\frac{\text{weight gain of rats fed seeds}}{\text{weight gain of rats fed casein}} \times 100.$

‡ Heated in autoclave at 121 °C for 15 min.

one of its N derivatives are responsible for most of the activity. The Swedish report (Agren & Lieden, 1968) does not identify the trypsin inhibitor but gives as evidence the enhanced PER of the heat-coagulated concentrate as compared to the aqueous 'porridge' of the flour before coagulation.

Uses of concentrates and isolates

Decorticated sunflower meal has potential as a human food because of its creamy color, pleasant nutty flavor and stability (Talley *et al.*, 1970; Kilara *et al.*, 1972; Robertson, 1972). The use of sunflower kernels, meal, flour and isolates in foods has been throughly discussed (Talley *et al.*, 1970; Burns *et al.*, 1972). Some of the many food uses of sunflower are (1) the kernel may be eaten as is or used as a nut substitute in bakery and confectionery products; (2) dehulled meal can be incorporated into breads and tortilla chips; (3) the flour can be used in batters and doughs; and (4) textured vegetable protein can be prepared from a protein isolate (McGregor, 1970; Talley *et al.*, 1970; Gandy, 1972; Robertson *et al.*, 1972). When sunflower flour was used in bread baking it caused an increase in water absorption of the dough, markedly influenced baking characteristics and caused a decrease in loaf volume. Sunflower flour was less compatible with wheat flour than were other oilseed flours (Rooney *et al.*, 1972), and imparted a grayish cast to light-colored products. Light-colored, isolates free of chlorogenic acid would alleviate this problem.

Several functional properties of sunflower protein have been studied. The fat and water binding characteristics and 'whipability' of sunflower meal are reported to compare favorably with soy meal (Huffman *et al.*, 1973). Factors such as ionic strength, method of extracting the kernel and isoelectric precipitation of the isolate markedly influence the N solubility of the meal and isolate respectively, whereas the presence of chlorogenic acid and isopropanol extraction of the isolate have little influence (Figs. 10.1 & 10.2). Solubility profiles often give an indication of the extent of protein denaturation.

Safflower

Safflower (*Carthamus tinctorius* L.) is one of the oldest cultivated oilseeds. It was found in the tombs of the Pharaohs and has been grown for centuries in the Nile Valley, the middle East and the Orient (Kneeland, 1958; Kohler, 1966). It was grown for the dyestuff carthamin which was used to color foods and fabrics.

The safflower plant is a thistle-like annual which grows from 0.5–1.5 m in height and has yellow to deep orange flower heads, each of which may contain 15–100 seeds. It grows best in arid climates with an adequate amount of moisture (Knowles, 1955; Kneeland, 1966; Kohler, 1966).

Production and yield

The production, processing and utilization of safflower have been thoroughly reviewed (Knowles, 1955; Kneeland, 1958, 1966; Waiss, 1971). A bibliography relating to safflower has also been compiled (Larson, 1962). The world production of safflower increased sharply in the early 1960s and has since stabilized (Table 10.5; Waiss, 1971). The rise in production reflects the increased use of the highly unsaturated safflower oil as edible vegetable oil rather than as an industrial oil (Kohler, 1966; Waiss, 1971). India and the USA lead the world in

Table 10.5. *Estimated world availability and exports of safflower seed (tonnes × 10³)*

	1955–9†	1964	1965	1966	1967	1968	1969	1970‡
World availability	100	576	502	527	594	463	376	571
World exports	34	218	171	196	209	127	100	190

† Average. ‡ Provisional.

Table 10.6. *Yield of safflower seed and protein in various regions†*

Region and conditions	Safflower seed (kg/ha)	Safflower protein‡ (kg/ha)	Reference
India, various regions	250–1750	35–230	Waiss (1971)
USA			
dry farming	250–800	35–100	Waiss (1971)
irrigation	1000–2000	130–260	
California	1880	245	Kneeland (1958)
USA			
dry farming	400–1600	50–210	Kohler (1966)
irrigation	2000–3300	260–430	
California	3300	435	Miller (1969)
Arizona	2650	345	Miller (1969)

† Values recalculated from lbs or tons/acre. ‡ Calculated as 13 % of safflower seed.

the production of safflower with annual crops of approximately 200000 t; Mexico follows closely with 180000 t (Waiss, 1971).

The yield of safflower seed is primarily a function of the quantity of available water, although environmental factors such as soil type and weather also have some influence (Kneeland, 1958). The range and variations of yields and estimates of these are summarized in Table 10.6. The effect of irrigation is very apparent.

Composition

The composition of safflower seed has been altered by the development of new thin-hulled and high oleic acid varieties. Older varieties consisted of 48–51 % kernel (where most of the protein and oil is contained) and 49–52 % hull or pericarp (which is mainly fiber). Newer thin-hulled varieties are 55–70 % kernel, the remainder being hull (Kneeland, 1958; Kohler, 1966; Waiss, 1971). Expressed in terms of the total seed, the newer varieties may contain 35–50 % oil, 13–17 % protein and 35–45 % hull (Smith, 1971; Waiss, 1971).

Table 10.7. *Composition of safflower seed on a % dry weight basis*

	Crude fat	Crude protein (N × 6.25)	Crude fiber	Ash	Nitrogen free extract
Whole seed	36.8–41.0	15.4–19.4	20.8–22.3	2.3–2.6	19–21.4
Hull	1.4–3.2	3.8–5.0	57.1–60.4	1.4–2.2	30.6–33.4
Kernel	59–64	24.9–29.4	1.0–1.6	2.6–3.9	6.9–9.5
Oil free whole seed	—	25.7–30.7	34.3–35.9	3.7–4.2	30–35.8
Oil free kernel	—	64.0–71.7	2.8–4.1	7.3–7.9	16.9–24.3

Data is from 4 commercial varieties; Gila, U-5, US-10 and Frio.

Table 10.8. *Composition of low-fiber products from brown-striped safflower seed on a % dry weight basis*

Product	Protein	Hull†	Fat	Ash	Other
Meal	64	9	1	8	18
Flour	67	6	1	8	17
Concentrate	81	7	1	9	2

† Calculated as crude fiber × 2.

The composition of the various parts of safflower seed is summarized in Table 10.7. The high fiber, and low protein content of the hull is very apparent.

Preparation of concentrates and isolates

One of the major benefits of concentrating or isolating safflower protein is the removal of fiber. Kopas & Kneeland (1966) described a process for preparing a high protein (41–45 %) meal which consisted of impacting a meal and classifying the material on the basis of particle size. The product is used as a feed supplement.

A low fiber, high protein flour and an edible concentrate were prepared by screening and aspirating cracked safflower seeds, extracting the oil, milling and sieving (Goodban, 1967). A light-colored, bitter-flavored flour was obtained which contained more fiber than the recommended 5 % maximum for human consumption. The edible concentrate was prepared by extracting the bitter and cathartic constituents from the flour with 70–80 % ethanol (Goodban, 1967; Kohler, 1969).

The bitter flavor of safflower is well recognized (Kohler, 1966; Palter & Lundin, 1970; Smith, 1971). Recently a lignan glycoside,

1-matairesinol-mono-β-D-glucoside was identified as a bitter principle in safflower meal (Palter & Lundin, 1970). Also, a tasteless lignan glycoside, 2-hydroxy-arctiin, was associated with cathartic activity (Palter *et al.*, 1972). The removal of these glycosides is imperative in the preparation of edible concentrates.

A process to obtain high protein, low fiber products from whole safflower seeds, press-cake meal or refined meal has been patented (Goodban & Kohler, 1970). Briefly, the process, developed on a pilot plant scale, involves heating safflower seeds to 60–65 °C, passing them through rollers to crush the kernels, carrying out a hexane extraction via counter current flow, collecting the suspended proteinaceous material by filtration or centrifugation and drying. The 1800 kg of whole safflower seed used yielded 180 kg of white flour-like material (65 % protein, 2.2 % fiber) suitable for animal feed. If isopropanol is used as the extractant an edible protein supplement is prepared.

Safflower protein may also be isolated by acid precipitation. Van Etten *et al.* (1963) initially studied the extractability of N from hexane-extracted kernels. They found that more than 90 % of the N was extracted at pH 9.0 and above, whereas pH 4.0 was the point of minimum extraction (≃ 10 %). Thus, kernels were extracted at pH 9.0, acidified to pH 4.0 to precipitate the protein, washed twice with water and dried at 60 °C. A yield of 45 g of the dark gray isolate (84 % protein) was obtained from 100 g of kernels. Additional extraction and/or purification to remove undesirable flavor and color constituents would be necessary before the isolate could be used as an edible product.

Elmquist (1965) described a more elaborate method for isolating safflower protein which involved extracting, coagulating, and spinning protein fiber. Again, it seems imperative that antinutritional and bitter components be removed from the isolate if these fibers are to be used in the human diet.

Nutritional evaluation

Amino acid analyses of safflower protein show lysine to be the first limiting amino acid with methionine and isoleucine present in less than optimal quantities (Table 10.9). It is interesting to note that the isolated protein of Van Etten *et al.* (1963) contained somewhat less lysine than the defatted kernel. A world collection of safflower was screened for high lysine seeds (Palter *et al.*, 1969), but no strikingly high mutants were found. The major difference between the meal and protein con-

Table 10.9. *Amino acid composition of safflower protein (g/16 g N)*

		Safflower meal					
Amino acid	FAO reference protein[a]	Partially decorticated, commercial[b]	Undecorticated brown, striped hull, experimental[b]	Lyman *et al.* (1956)	Safflower kernel P-1[c]	Isolated safflower protein P-1[c]	Soybean[d]
Arginine	—	8.7	8.3	7.8[3]	9.4	9.7	6.9
Histidine	—	2.3	2.2	2.0	2.6	2.6	2.5
Isoleucine	4.2	4.0	3.8	3.8	3.7	4.4	4.8
Leucine	4.8	6.1	5.9	5.5	6.0	6.8	7.3
Lysine	4.2	2.8	2.6	2.7	3.2	2.9	5.8
Methionine	2.2	1.6	1.4	1.5	1.5	1.6	1.4
Phenylalanine	2.8	4.3	4.3	5.2	4.3	5.1	4.8
Threonine	2.8	3.1	2.9	2.9	3.2	3.1	3.8
Tryptophan	1.4	—	—	1.2	0.9	—	1.7
Valine	4.2	5.5	5.3	4.9	5.3	6.2	5.0

[a] FAO (1965); [b] Guggolz *et al.* (1968); [c] van Etten *et al.* (1963); [d] Kohler *et al.* (1966).

centrate, i.e., fiber, must be kept in mind as the data are presented. Animal studies carried out support several general conclusions: (1) lysine is the first limiting amino acid; methionine and isoleucine are also low (Kohler, 1966; Evans & Bandemer, 1967; Kuzmicky, 1967); (2) chicks grow better on adequately supplemented safflower meal than on methionine supplemented soybean meal (Fisher *et al.*, 1962; Kohler *et al.*, 1966); (3) safflower protein is most successfully utilized in conjunction with high lysine protein, e.g., soybean, fish or meat meal (Kneeland, 1958; Kuzmicky & Kohler, 1968*b*); and (4) safflower meal rations provide poorer feed efficiency than soybean meal due to the lower metabolizible energy (Kohler *et al.*, 1966; Kuzmicky, 1967). There is little information on the biological value (BV), total digestibility (TD), or protein efficiency ratios (PER) of safflower protein. Coefficients of TD of 86.5–92.4 and BVs of 84.9 and 86.0 have been reported (Kohler, 1966; Shoji *et al.*, 1966). Although the PER of safflower flour (58 % protein) was only 1.39, it was increased to 2.09 by the addition of lysine and methionine (Kohler, 1966). In terms of protein nutritive value (PNV), there was great variability among varieties of ground, hexane-extracted safflower seeds (Table 10.4; Evans & Bandemer, 1967). Although Gila contained more lysine than US-10, it had to be supplemented with both lysine and methionine before its PNV

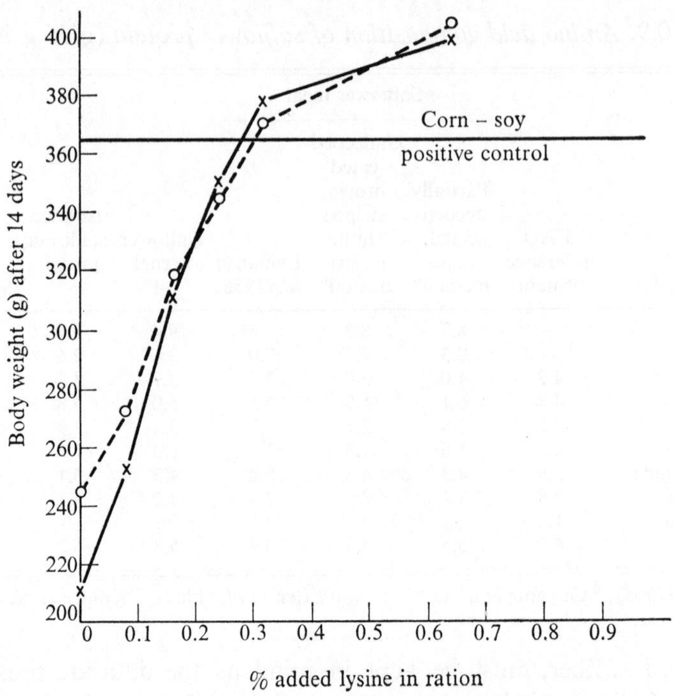

Fig. 10.3. Lysine dose response curves over a 2 week period for chicks on safflower-meal-based rations supplemented with lysine. ×—×, corn+safflower ration; ○--○, glucose+safflower ration. Mean starting weight of chicks at 5 days was 87 g.

was equivalent to that of US-10. There are apparent differences in lysine availability amongst safflower varieties which should be studied.

The use of safflower protein as a supplement has been reviewed (Kneeland, 1958; Kuzmicky & Kohler, 1968*b*; Waiss, 1971). Safflower meal has been used effectively in the diets of chicks and laying hens (Kratzer, 1951; Peterson *et al.*, 1957; Young & Halloran, 1962; Valadez *et al.*, 1965; Kuzmicky & Kohler, 1968*b*), pigs (Waiss, 1971), chinchillas and rabbits (Kneeland, 1958) as well as lambs and cattle (Baker *et al.*, 1951, 1960; Goss & Otagaki, 1954). Safflower was invariably either combined with a high lysine protein or supplemented with lysine and usually provided 20–50 % of the total protein in the ration. In several experiments, lysine and methionine supplemented safflower rations promoted chick growth better than methionine supplemented soy or corn–soy diets (Kohler *et al.*, 1966; Kuzmicky & Kohler, 1968*a*). The influence of lysine on the growth response of chicks fed a corn–safflower ration is evident in Fig. 10.3. Fig. 10.4 provides a dramatic

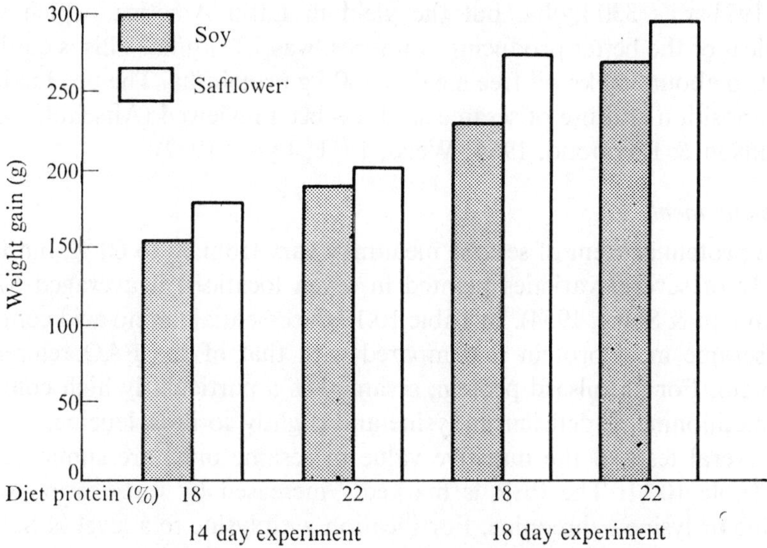

Fig. 10.4. Comparison of weight gains of chicks fed safflower or soybean meal as the main source of protein.

comparison of adequately supplemented safflower and soybean rations with the 18 day diets being isocaloric (Kohler *et al.*, 1966). These studies show that adequately supplemented safflower protein has good nutritional potential.

Uses of concentrates and isolates

The potential of safflower protein as a human food was reviewed by Kohler (1966). There are, however, few reports on its functionality and/or incorporation into foods. An edible safflower flour (70 % protein) was incorporated into meat-like patties and bread at a level of 5 % of the flour. Loaf volume decreased slightly and there was a distinct though not undesirable flavor present. The fibers described by Elmquist (1965) have potential as simulated fibrous foods, e.g., meat- and fishlike products.

Sesame

Sesame (*Sesamum indicum*) is one of the most ancient, widespread and valuable oilseeds. The seed contains about 50 % of a particularly stable oil with a desirable flavor and about 25 % protein that has a favorable balance of amino acids and a particularly high content of methionine. World production in 1971 was 2.1 Mt (FAO, 1971). The average yield in 1971 was 330 kg/ha, but the yield in Latin America, which was typical of the better producing countries, was 720 kg/ha. This is equivalent to about 360 kg oil-free meal or 180 kg protein/ha. The production, composition and use of sesame seed has been reviewed (Altschul, 1958; Johnson & Raymond, 1964; Weiss, 1971; Lyon, 1972).

Sesame meal

The protein content of sesame meal may vary from 30 to 60 %, but in a study of several varieties planted in seven locations it averaged 57 % (Kinman & Stark, 1954). In Table 10.10 the essential amino acid content of sesame meal protein is compared with that of the FAO reference protein. For an oilseed protein, sesame has a particularly high content of methionine, is deficient in lysine and slightly so in isoleucine.

Several tests of the nutritive value of sesame meal are summarized in Table 10.11. The PER is markedly increased by fortification with lysine or lysine + threonine. Fortification with lysine to a level of 8.2 %, the level in casein, raised the PER to 2.91, as high as that of casein. The addition of sesame meal, as 25–50 % of the feed protein, to corn meal, wheat flour and peanut meal + Bengal gram rations markedly increased the PER. In a similar rat feeding study (Evans & Bandemer, 1967), the protein nutritive value of sesame meal was increased twofold, i.e., up to that of casein, by supplementation with 0.2 % lysine + 0.1 % isoleucine + 0.1 % methionine or by mixing with an equal weight of soybean meal, which has a high lysine content.

In poultry feeds, supplementation of sesame meal with 0.4–0.5 % lysine or with soybean meal was found to be necessary for maximum growth and feed efficiency (Cuca & Sunde, 1967). In human feeding tests, a sesame + peanut + Bengal gram protein mixture, which is somewhat deficient in lysine, was nearly as effective as skim milk in controlling the clinical manifestations of protein malnutrition but was inferior to skim milk with regard to serum albumin regeneration (Gopalan, 1961).

Sesame seeds contain 1–2 % oxalic acid in the thin hull. To obtain a non-bitter, low-fiber, higher protein meal, this hull should be removed. Seeds may be decorticated by treatment with alkali (Horvilleur, 1957; Lucidi, 1962; Shamanthaka Sastry *et al.*, 1969) before conventional processing. Hulls have also been removed from solvent-extracted meals by screening or air classification (Carter *et al.*, 1961). Protein content of the meals was raised from 51–6 % before screening to 60–8 %, and to 80 % when the seed was decorticated before solvent extraction.

Table 10.10. *Essential amino acid content of sesame protein (g/16 g N)*

Amino acid	Sesame protein†	FAO reference protein
Arginine	12.0–13.0	2.0
Histidine	2.4–2.8	2.4
Isoleucine	3.3–3.6	4.2
Leucine	6.5–7.0	4.8
Lysine	2.5–3.0	4.2
Methionine	2.5–4.0	2.2
Methionine+cystine	3.8–5.5	4.2
Phenylalanine	4.2–4.5	2.8
Threonine	3.4–3.8	2.6
Tryptophan	2.0–2.4	1.4
Valine	4.2–4.4	4.2

† Average of 5 varieties (Evans & Bandemer, 1967).

Table 10.11. *Efficiency of protein utilization with feeds containing sesame meal. Data from rat feeding tests*

Protein source	% protein in ration	PER†
Sesame meal[a]	9	1.35
sesame meal+0.2 % lysine	9	1.57
sesame meal+0.2 % lysine+0.2 % threonine	9	1.82
Sesame meal[b]	10	1.70
sesame meal fortified to lysine content of 4.2 % of amino acids	10	2.14
sesame meal fortified to lysine content of 8.2 % of amino acids	10	2.91
White corn meal	6.9	0.90
white corn meal (87 %)+sesame meal (13 %)[c]	6.9	1.81
Enriched wheat flour	9	1.08
enriched wheat flour (91 %)+sesame meal (9 %)[d]	9	1.17
Peanut meal+Bengal gram[e]	10	1.79
peanut meal+Bengal gram+sesame meal[f]	10	2.03

† Protein efficiency ratio = g gain per g protein eaten. [a] 10 week test (Kik, 1960); [b] 4 week test (Joseph *et al.*, 1962); [c] 10 week test, protein in ration: 50 % from corn, 50 % from sesame (Kik, 1960); [d] 10 week test, protein in ration: 66.7 % from wheat, 33.3 % from sesame (Kik, 1960); [e] 4 week test, protein in ration: 50 % from peanut, 50 % from Bengal gram (Joseph, 1958); [f] 4 week test, protein in ration: 50 % from peanut, 25 % from Bengal gram, 25 % from sesame (Joseph, 1958).

Sesame protein isolates and hydrolysates

Sesame protein has been extracted from the meal with several solutions and at varying pH, as summarized in Table 10.12. Generally, extraction with alkali at pH 9–11 is most effective. Protein isolates have been precipitated from these extracts by acidification with sulfuric acid to a pH

Table 10.12. *Extraction of protein from sesame meal*

Extracting solvent	pH	% protein extracted	Precipitation of protein, pH
0.5 % NaOH[a]	11.7	76	—
NaOH[b]	10.5	—	4.0
NaOH[c]	10.0	—	4.8
NaOH[d]	9.0	87	—
0.4 % Na_2SO_3[a]	8.2	49	—
0.6 % $NaHSO_3$[d]	—	54	—
10 % NaCl[e]	—	80	—
6 % NaCl[d]	—	75	—
HCl[d]	1.3	36	—

[a] Kodangekar (1946); [b] Arthur & Volkert (1950) 50 % yield of precipitated protein; [c] Rastogi & Krishna Murti (1962); [d] Basu & Sen Gupta (1947); [e] Nath & Giri (1957).

of 4.0 and 4.8. Protein isolate blends with desirable contents of essential amino acids have been prepared by extracting mixtures of sesame and soy meals with NaOH plus a little H_2O_2 at pH 10.5, separating the extract and precipitating the protein at pH 4.5 with HCl (Johnson & Anderson, 1968).

Sesame meal has been partially hydrolyzed by treatment with the fungal protease from *Trametes sanguina* (Sreekantiah *et al.*, 1969). The meal was steamed for 15 min, acidified to pH 2.5–3.0 and treated for 5 h at 45 °C with 0.25 % enzyme. The freeze-dried extract contains 64 % soluble protein as against 7.2 % soluble protein and 49 % total protein in the meal. The hydrolysate showed increases in all the amino acids except methionine.

The nutritive value of sesame meal, protein isolate and protein hydrolysate have been compared (Rastogi & Krishna Murti, 1962). The hydrolysate, prepared by treatment with papain, extracted 50 % of the meal protein. Biological values, estimated by N balance on adult rats, were sesame meal (70 %), protein isolate (59 %), protein hydrolysate (70 %) and casein (62 %). Protein efficiency ratios of casein and the hydrolysate were similar; values for the meal and isolate were lower.

Castor

Castor (*Ricinus communis*) is widely cultivated for its oil. The seed contains about 50 % oil and 18 % protein. The unique oil, in which ricinoleic acid comprises 90 % of its fatty acids, is an important industrial raw material. During 1967–71, annual production of castor averaged 857000 t (FAO, 1971). Average yield in Brazil, the largest producer, was 970 kg/ha. This is equivalent to 485 kg oil-free meal or 175 kg protein/ha. Processing, composition and use of castor meal have been reviewed by Altschul (1958), Fuller *et al.* (1971) and Waiss (1971).

Castor meal

The meal or pomace remaining after expression or extraction of the oil presents many problems because of its toxic components (Jenkins, 1963). It contains a toxic protein, ricin, some potent protein–polysaccharide allergens (Spies & Coulson, 1943), and 0.3–0.4 % of an alkaloid, ricinine. Because of these toxic components castor meal is not used as a food protein, but is used largely as a fertilizer. However, meal that has been detoxified, and preferably deallergenized, has a potential for use in animal feeds.

Ricin is readily detoxified by heating with steam (Kodras *et al.*, 1949), and commercial castor meal can be detoxified by the introduction of about 20 % moisture during heating to remove the solvent after oil extraction (Bris & Algeo, 1970; Fuller *et al.*, 1971).

However, the castor allergens are not destroyed by normal detoxification and people handling this meal may become sensitized and suffer allergic reactions. Recently, pilot plant processes have been developed using steam, lime and ammonia treatments (Fuller *et al.*, 1971; Mottola *et al.*, 1971; Mottola *et al.*, 1972). Optimum inactivation of the allergens was accomplished by treatment with steam for 1 h at 0.7 kg/cm^2 gauge pressure, 6 N ammonia for 45 min at 80 °C or 4 % lime for 15 min at 100 °C. Though most of the allergen was inactivated, the content of some of the essential amino acids was reduced, particularly by lime treatment.

Ricinine in diets with a high proportion of castor meal inhibits the growth of chicks (Murase *et al.*, 1966); for maximum feed efficiency with chicks the ricinine level should not be above that obtained when using castor meal as 10–15 % of the rations (Fuller *et al.*, 1971).

The growth-depressing component, probably ricinine, of detoxified castor meal has been extracted with hot water (Vilhjalmsdottir & Fisher, 1971). The commercial meal used was detoxified by steaming for 4 h at 5.4 kg/cm^2, and contained 42.5 % protein with the amino acid analysis listed in Table 10.13. Hot water extraction caused no significant change in protein or amino acid content, but improved the growth response of chicks. Chick feeding studies showed that lysine was the first and tryptophan the second limiting amino acid. The extracted meal,

Table 10.13. *Amino acid composition of castor meal protein*

Amino acid	Amount (g/16 g N)	Amino acid	Amount (g/16 g N)
Valine	5.44	Tryptophan	0.31
Isoleucine	4.68	Tryosine	2.82
Leucine	6.42	Glycine	4.31
Threonine	3.44	Alanine	4.26
Methionine	1.51	Proline	3.74
Cystine/2	0.68	Serine	5.44
Phenylalanine	4.02	Hydroxyproline	0.28
Lysine	2.68	Aspartic acid	9.67
Histidine	1.25	Glutamic acid	18.87
Arginine	8.61		

Table 10.14. *Growth and protein utilization by chicks fed water-extracted, detoxified castor meal*

Protein (17% of ration)	Amino acid supplement	Body weight† (g)	Feed utilization (g gain/g feed)	Net protein utilization (%)
Castor meal	—	67	0.03	25
Castor meal	0.8% lysine +0.2% tryptophan +0.2% methionine‡	145	0.37	45
Soy protein isolate	0.2% methionine	140	0.44	53

† 7-day-old chick fed 10 days. ‡ Equivalent growth obtained without methionine.

supplemented with lysine and tryptophan, gave growth rates of chicks as good as those obtained with methionine supplemented soy protein (Table 10.14).

Several beef-cattle feeding tests with detoxified castor meal have been summarized recently (Bris & Algeo, 1970; Weiss, 1971). The cattle showed good weight gains, produced high quality beef, and no toxic effects were evident in over 300 animals fed in several experiments at different locations. A net energy difference between cottonseed meal and castor meal could be made up with a small amount of residual castor oil in the meal.

Commercial, partially detoxified castor meal has been fed to a group of 10 dairy cattle for one year (Walker *et al.*, 1972). All animals met their previous lactation levels and no toxic effects were evident. Transmission of allergen was very slight. A trace of ricin appeared in the milk in only one of dozens of samples. Hydroxy-fatty acids appeared in milk at the limit of detection when 0.5 % castor oil was added to the diet. Milk from these tests was lyophilized and fed to rats at a 20 % level for 98 days. Growth, development and blood and urine analyses were completely normal.

Castor protein isolates and hydrolysates

Castor meal protein has been extracted with 0.2–0.4 % NaOH, 0.3 % Na_2SO_3 and 10 % NaCl (Kodangekar *et al.*, 1946; Vassel, 1952; Joshi & Varma, 1955). Toxicity and nutritional properties of these isolated proteins were not investigated.

Castor meal has been fermented with a proteolytic bacterial culture, preferably *Clostridium*, then autoclaved and dried to yield a material containing 32 % partly digested protein that was claimed to be non-toxic and have a high feed value for mammals and fowl (Darzins, 1960).

References

Agren, G. & Lieden, S.-A. (1968). Some chemical and biological properties of a protein concentrate from sunflower seeds. *Acta Chemica Scandinavica*, **22**, 1981.

Altschul, A. M. (1958). *Processed plant protein foodstuffs* (ed. A. M. Altschul). Academic Press, New York.

Arthur, J. C. & Volkert, E. C. (1950). Physical and chemical properties of sesame protein. *Journal of Southern Research*, **2**, 5.

Baker, G. N., Böksu, T. & Baker, M. L. (1960). Undecorticated safflower as a protein supplement for wintering calves. Nebraska Agricultural Experiment Station Bulletin No. 458.

Baker, M. L., Baker, G. N., Erwin, C., Harris, L. C. & Alexander, M. S. (1951). Feeding safflower meal. Nebraska Agricultural Experiment Station Bulletin No. 402.

Bandemer, S. L. & Evans, R. T. (1963). The amino acid composition of some seeds. *Journal of Agricultural and Food Chemistry*, **11**, 134.

Basu, U. P. & Sen Gupta, S. K. (1947). Extraction of protein from sesame cake. *Indian Journal of Pharmacology*, **9**, 60.

Bris, E. J. & Algeo, J. W. (1970). Castor seed byproducts for cattle rations. *Feedstuffs*, **42**, 26.

Brummett, B. T. & Burns, E. E. (1972). Pigment and chromogen characteristics of sunflower seed. *Journal of Food Science*, **37**, 1.

Burns, E. E., Talley, L. T. & Brummett, B. T. (1972). Sunflower utilization in human foods. *Cereal Science Today*, **17**, 287.

Cancalon. P. (1971). Chemical composition of sunflower seed hulls. *Journal of the American Oil Chemists' Society*, **48**, 629.

Carter, R. L., Cirino, V. O. & Allen, L. E. (1961). Effect of processing on composition of sesame seed and meal. *Journal of the American Oil Chemists' Society*, **38,** 148.

Cater, C. M., Gheyasuddin, S. & Mattil, K. F. (1972). The effect of chlorogenic, quinic and caffeic acids on the solubility and color of protein isolates, especially from sunflower seed. *Cereal Chemistry*, **49**, 508.

Clandinin, D. R. (1958). Sunflower seed oil meal. In: *Processing plant protein foodstuffs* (ed. A. A. Altschul), p. 557. Academic Press, New York.

Cuca, M. & Sunde, M. L. (1967). Amino acid supplementation of a sesame meal diet. *Poultry Science*, **46,** 1512.

Darzins, E. (1960). Edible castor cake. US Patent No. 2920963.

Earle, F. R., Van Etten, C. H., Clark, T. F. & Wolf, I. A. (1968). Compositional data on sunflower seed. *Journal of the American Oil Chemists' Society*, **45**, 876.

Elmquist, L. F. (1965), US Patent No. 3175909 (March 30).

Evans, R. J. & Bandemer, S. L. (1967). Nutritive value of some oilseed proteins. *Cereal Chemistry*, **44**, 417.

FAO (1965). Nutrition Meeting Report Series, No. 37. United Nations, Rome.

FAO (1971). *Production Yearbook* 25, pp. 250–9, United Nations, Rome.

Feeny, P. P. (1968). Inhibitory effect of oak leaf tannins on the hydrolysis of proteins by trypsin. *Phytochemistry* **8,** 2119.

Fisher, H., Summers, T. D., Wessels, T. P. H. & Shapiro, R. (1962). Further evaluation of protein for the growing chicken by carcass retention method. *Journal of the Science of Food and Agriculture*, **13**, 658.

Fuller, G., Walker, H. G., Mottola, A. C., Kuzmicky, D. D., Kohler, G. O. & Vohra, P. (1971). Potential for detoxified castor meal. *Journal of the American Oil Chemists' Society*, **48**, 616.

Gandy, D. E. (1972). Projection and prospects for sunflower seed. *Journal of the American Oil Chemists' Society*, **49**, 518A.

Gheyasuddin, S., Cater, C. M. & Mattil, K. F. (1970*a*). Preparation of a colorless sunflower protein isolate. *Food Technology*, **24**, 242.

Gheyasuddin, S., Cater, C. M. & Mattil, K. F. (1970*b*). Effect of several variables on the extractability of sunflower proteins. *Journal of Food Science*, **35**, 453.

Girault, A., Baudet, T. & Mosse, T. (1970). Improvement of the lysine content of the protein of sunflower seeds. *Improvement of Plant Protein by Nuclear Technology*. Symposium Proceedings, p. 275.

Glicksman, M. (1971). Fabricated foods. *Critical Review of Food Technology*, **2**, 21.

Goodban, A. E. (1967). Safflower protein products for food use. In: First Research Conference on the Utilization of Safflower, May 25–6, 1967, Albany, California.

Goodban, A. E. & Kohler, G. O. (1970). US Patent No. 3542559 (November 24).

Gopalan, C. (1961). Protein malnutrition in India. In: *Meeting protein needs of infants and children* (ed. L. Voris), p. 211. National Academy of Sciences/National Research Council, Washington DC.

Goss, H. & Otagaki, H. H. (1954). Safflower meal digestion tests. *California Agriculture*, **8**, 15.

Guggolz, J., Rubis, D. D., Herring, V. V., Palter, R. & Kohler, G. O. (1968). Composition of several types of safflower seed. *Journal of the American Oil Chemists' Society*, **45**, 689.

Horvilleur, G. (1957). Sesame seed decortication. US Patent No. 2815783.

Huffman, L. M., Brummett, B. J. & Burns, E. E. (1973). Sunflowers as food. In *League for International Food Education Newsletter*, March 1973 (ed. S. M. Weisberg). Washington.

Jenkins, F. P. (1963). Allergenic and toxic components of castor meal: review of the literature and studies of the inactivation of these components. *Journal of the Science of Food and Agriculture*, **14**, 773.

Johnson, R. A. & Anderson, P. T. (1968). Blended protein isolation process. US Patent No. 3397991.

Johnson, R. H. & Raymond, W. D. (1964). Chemical composition of some tropical food plants. III Sesame seed. *Tropical Science*, **6**, 173.

Joseph, A. A., Tasker, P. K., Joseph, K., Rao, N. N., Swaminathan, M., Sankaran, A. N., Sreenivassan, A. & Subrahmanyan, V. (1962). Fortification of sesame meal with lysine. *Annals of Biochemistry and Experimental Medicine (Calcutta)*, **22**, 113.

Joseph, K. (1958). Supplementary values of the proteins of Bengal gram and sesame to groundnut protein. *Food Science*, **7**, 186.

Joshi, B. N. & Varma, J. P. (1955). Castor seed proteins and their viscosities. *Journal of the American Oil Chemists' Society*, **32**, 553.

Kik, M. C. (1960). Effect of amino acid supplements on the nutritive value of proteins in sesame seed and meal. *Journal of Agricultural and Food Chemistry*, **8**, 327.

Kilara, A., Humbert, E. S. & Sosulski, F. W. (1972). Nitrogen extractability and moisture adsorption characteristics of sunflower seed products. *Journal of Food Science*, **34**, 771.

Kinman, M. L. & Stark, S. M. (1954). Yield and composition of sesame as affected by variety and location grown. *Journal of the American Oil Chemists' Society*, **31**, 104.

Kneeland, J. A. (1958). Minor oilseeds and tree nut meals. In: *Processed plant protein foodstuffs* (ed. A. A. Altschul) p. 619. Academic Press, New York.

Kneeland, J. A. (1966). The status of safflower. *Journal of the American Oil Chemists' Society*, **43**, 403.

Knowles, P. F. (1955). Safflower – production, processing and utilization. *Economic Botany*, **9**, 273.

Kodangekar, P. R., Mandlekar, M. R. & Mehta, T. N. (1946). Utilization of vegetable oil cakes for the production of protein. *Journal of the Indian Chemical Society* (Ind. & News Ed.), **9**, 34.

Kodras, R., Whitehair, C. K. & MacVicar, R. (1949). Studies on the detoxication of castor meal pomace. *Journal of the American Oil Chemists' Society*, **26**, 241.

Kohler, G. O. (1966). Safflower, a potential source of protein for human food. In: *World Protein Resources* (ed. R. F. Gould). *Advances in Chemistry Series*, **57**, 243.

Kohler, G. O. (1969). Utilization prospects of safflower. In: Third Safflower Research Conference, May 7–8, 1969, Davis, California.

Kohler, G. O., Kuzmicky, D. D., Palter, R., Guggolz, J. & Herring, V. V. (1966). Safflower meal. *Journal of the American Oil Chemists' Society*, **43**, 413.

Kopas, G. A. & Kneeland, J. A. (1966). US Patent No. 3 271 160 (September 6).

Kratzer, F. H. & Williams, D. (1951). Safflower oil meal in rations for chicks. *Poultry Science*, **30**, 417.

Kromer, G. W. (1967). Sunflowers gain as an oilseed crop in the United States. ERS-360, Economic and statistical analysis division, Economic Research Service, Washington DC.

Kuzmicky, D. D. (1967). Safflower meal in poultry rations. In: First Research Conference on Utilization of Safflower, May 25–6, 1967, Albany, California.

Kuzmicky, D. D. & Kohler, G. O. (1968*a*). Safflower meal – utilization as a protein source for broiler rations. *Poultry Science*, **47**, 1266.

Kuzmicky, D. D. & Kohler, G. O. (1968*b*). Safflower meal – the effect of chick age and ration lysine content on its use in chick starter rations. *Poultry Science*, **47**, 1473.

Larson, N. G. (1962). Safflower 1900–1960 a list of selected references. Library List 73, National Agricultural Library, Washington, DC.

Loomis, W. D. & Battaile, J. (1966). Plant phenolic compounds and the isolation of plant enzymes. *Phytochemistry*, **5**, 423.

Lucidi, I. C. (1962). Decorticating sesame seeds. US Patent No. 3054433.

Lyman, C. M., Kuiken, K. A. & Hale, F. (1956). Essential amino acid content of farm feeds. *Journal of Agricultural and Food Chemistry*, **4**, 1008.

Lyon, C. K. (1972). Sesame: composition and use. *Journal of the American Oil Chemists' Society*, **49**, 245.

McGregor, D. (1970). Formulation of new sunflower seed products. In: Proceedings of the 4th International Sunflower Conference, Memphis, Tennessee.

Milic, B., Stojanovic, S., Vulcurevic, N. & Turcis, M. (1968). Chlorogenic and quinic acids in sunflower meal. *Journal of the Science of Food and Agriculture*, **19**, 108.

Miller, M. D. (1969). Economic and utilization outlook. In: Third Safflower Research Conference, May 7–8, 1969, Davis, California.

Mottola, A. C., Mackey, B. & Herring, V. (1971). Castor meal antigen deactivation – pilot plant steam process. *Journal of the American Oil Chemists' Society*, **48**, 510.

Mottola, A. C., Mackey, B., Herring, V. & Kohler, G. (1972). Castor meal antigen deactivation – pilot plant ammonia process. *Journal of the American Oil Chemists' Society*, **49**, 101.

Murase, K., Kusakawa, S., Yamaguchi, C., Takahashi, T., Funatsu, M., Goto, I., Koya, O. & Okamato, S. (1966). Toxicity of ricinine in castor meal *Journal of the Agricultural Chemical Society of Japan*, **40**, 61.

Nath, R. & Giri, K. V. (1957). Solubilization of proteins of sesame and characterization by electrophoresis. *Journal of Scientific and Industrial Research*, C, **16**, 5.

O'Connor, D. E. (1971). US Patent 3622556. (November 13).

Ohlson, R. (1972). Projection and prospects for rapeseed and mustard seed. *Journal of the American Oil Chemists' Society*, **49**, 522A.

Palter, R., Kohler, G. O. & Knowles, P. F. (1969). Survey for a high lysine variety in the world collection of safflower. *Journal of Agricultural and Food Chemistry*, **17**, 1298.

Palter, R. & Lundin, R. E. (1970). A bitter principle of safflower: matairesinol monoglucoside. *Phytochemistry*, **9**, 2407.

Palter, R., Lundin, R. E. & Haddon, W. F. (1972). A cathartic lignan glycoside isolated from carthamus tinctorius. *Phytochemistry*, **11**, 2871.

Panchenko, A. Y. (1966). Sunflower production and breeding in the USSR. In: Second International Sunflower Conference, August 17–18, 1966, Manitoba.

Peterson, C. F., Wiese, A. C., Anderson, G. J. & Lampman, G. E. (1957). The use of safflower oil meal in poultry rations. *Poultry Science*, **36**, 3.

Pomenta, J. V. & Burns, E. E. (1971). Factors affecting chlorogenic, quinic and caffeic acid levels in sunflower seed kernels. *Journal of Food Science*, **36**, 490.

Pomenta, T. F. (1970). Chemical and physical characteristics of selected types of sunflower seeds. MS Thesis. Texas A & M University, College Station, Texas.

Posey, M. H. (1969). Sunflower: a literature survey. Jan. 1960–June 1967. Library List No. 95, National Agricultural Library, Washington, DC.

Rastogi, M. K. & Krishna Murti, C. R. (1962). Biological value of sesame cake, protein isolate and protein hydrolysate. *Annals of Biochemistry and Experimental Medicine* (*Calcutta*), **22**, 51.

Reifer, L. & Augustyniak. (1968). Preliminary identification of the arginase inhibitor from sunflower seed. *Bulletin de l'Academie polonaise des Sciences*, Cl. II, **16**, 139.

Robertson, J. A. (1972). Sunflower: America's neglected crop. *Journal of the American Oil Chemists' Society*, **49**, 239.

Robertson, J. A., Thomas, J. K. & Burdick, D. (1971). Chemical composition of the seed of sunflower hybrids and open pollinated varieties. *Journal of Food Science*, **36**, 873.

Rooney, L. W., Gustafson, C. B., Clark, S.P. & Cater, C. M. (1972). Comparison of the baking properties of several oilseed flours. *Journal of Food Science*, **37**, 14.

Rutkowski, A. (1971). The feed value of rapeseed meal. *Journal of the American Oil Chemists' Society*, **48**, 863.

Sechet-Sirat, T., Masqueitar, T. & Tayeay, F. (1959). Pigments in sunflower seeds. I. Chlorogenic acid. *Bulletin de la Société de chimie biologique*, **41**, 1059.

Shamanthaka Sastry, M. C., Subramanian, N. & Rajagopalan, R. (1969). Wet dehulling of sesame seed to obtain superior protein concentrates. *Journal of the American Oil Chemists' Society*, **46**, 592A.

Shoji, K., Tajima, M., Totsuka, K. & Iwai, H. (1966). Feeding value of safflower meal. *Japanese Poultry Science*, **3**, 63.

Smith, A. K. (1971). Practical consideration in commercial utilization of oilseeds. *Journal of the American Oil Chemists' Society*, **48**, 38.

Smith, K. J. (1968). A review of the nutritional value of sunflower meal. *Feedstuffs*, **40**, 20.

Sosulski, F. W. & Bakal, A. (1969). Isolated proteins from rapeseed, flax, and sunflower meals. *Journal of the Canadian Institute of Food Technology*, **2**, 28.

Sosulski, F. W., McClearly, C. W. & Soliman, F. S. (1972). Diffusion extraction of chlorogenic acid from sunflower kernels. *Journal of Food Science*, **37**, 253.

Spies, J. R. & Coulson, E. J. (1943). Isolation and properties of an active protein – polysaccharidic fraction from castor seeds. *Journal of the American Chemical Society*, **65**, 1720.

Sreekantiah, K. R., Ebine, H., Ohta, T. & Nakano, M. (1969). Enzymic processing of vegetable protein foods. *Food Technology*, **23**, 1055.

Talley, L. D., Brummett, B. J. C. & Burns, E. E. (1970) Utilization of sunflower in human food products. In: Proceedings of the 14th International Sunflower Conference, Memphis, Tennessee.

Trotter, W. K. & Givan, W. D. (1971). Economics of sunflower oil production and use in the United States. *Journal of the American Oil Chemists' Society*, **48**, 442A.

Valadez, S., Featherston, W. R. & Pickett, R. A. (1965). Utilization of safflower meal by the chick and its effect upon plasma lysine and methionine concentrations. *Poultry Science*, **44**, 909.

Van Etten, C. H., Rackis, J. J., Miller, R. W. & Smith, A. K. (1963). Amino acid composition of safflower kernels, kernel protein, and hulls and solubility of kernel nitrogen. *Journal of Agricultural and Food Chemistry*, **11**, 137.

Vassel, B. (1952). Extracting vegetable protein. US Patent No. 2607767.

Vilhjalmsdottir, L. & Fisher, H. (1971). Castor meal as a protein source for chickens: detoxification and determination of limiting amino acids. *Journal of Nutrition*, **101**, 1185.

Waiss, E.A. (1971). *Castor, sesame and safflower*. Barnes-Noble Inc., New York.

Walker, H. G., Robb, J. G., Laben, R. C. & Herring, V. (1972). Use of castor meal in dairy rations. In: Abstracts of the 57th annual meeeting of American Association of Cereal Chemists, Miami, 1972.

Wolf, W. J. (1972). What is soy protein? *Food Technology*, **26**, 44.

Young, R. D. & Halloran, H. R. (1962). Decorticated safflower meal in chicken rations. *Poultry Science*, **41**, 1696.

11. Groundnut

O. L. OKE, R. H. SMITH & A. A. WOODHAM

The groundnut (*Arachis hypogaea* Linn.) is an annual herb belonging to the division Papilionaceae of the family Leguminosae. Needing a hot climate for development, with moderate rainfall or irrigation and ample sunshine during the growing season, it is suitable for tropical, subtropical, mediterranean and warm-temperate climates. The different varieties can be broadly divided into two types – the bushy upright or erect, and the trailing or runner. The plant, which has pinnate leaves and yellow flowers, usually grows up to 25–50 cm high and spreads rapidly. When the corolla has dropped the flower bends down, the ovary elongates and the fruit penetrates the soil where the pod, containing about one to three seeds develop. The soil should be well-drained; preferably a fertile sandy loam.

Cultivation

The seeds are usually planted at the rate of about 37–50 kg of unhulled seeds or 24–37 kg of hulled seeds per hectare, spaced at about 45 × 30 cm, but considerable variations occur between countries. Increase in kernel yield which might be expected from an increase in the number of plants per hectare is offset by a reduction in the number of nuts per plant (Meredith, 1964). Germination occurs within 6–10 days for undecorticated seeds and 4–5 days for decorticated seeds.

Usually there is very little response to fertiliser nitrogen, though it has been found in North Africa and India that yield may be increased by about 20 % by the application of nitrogenous fertilisers at the rate of 25 kg/ha N. The application of 1–3 tons of farm yard manure per acre has been found to be beneficial (Oyenuga, 1967). In Nigeria the application of 0.5–1.0 cwt. superphosphate per acre has been found to increase yields and this is attributed to the calcium and sulphur present which aid nodulation (Greenwood, 1949, 1951).

The growing season is normally 120–140 days but early varieties may allow a second crop in some areas. The seeds are usually harvested when the foliage begins to go light yellow, and delay should be avoided because of the possibility of germination of the seed. After sun-drying, which should not be prolonged because of a danger of the seed splitting, the seeds are removed by threshing and stored in a cool dry place. Improper storage may lead to the absorption of foreign flavours and may also result in contamination with *Aspergillus flavus*, the fungus responsible for the production of aflatoxin. The optimum moisture content for storage appears to be about 6.5 %.

The nutritive value of groundnuts

Groundnuts have been propagated for use as food for at least 600 years and the uncultivated natural species must have been consumed by man and animals even before that. It is one of the most concentrated foods because of its high fat and protein content. Thus 1 g provides 25 kJ compared with pure sugar (17 kJ), polished rice (15 kJ) and maize (14.7 kJ) (Oyenuga, 1968). Because of its high protein content, it could satisfy an appreciable proportion of the daily protein requirement if consumed in the countries of origin. Even in India, despite the large population, it has been estimated that groundnut could provide about 10 % of the protein, assuming that 40–70 g of crude protein (CP = N × 6.25) is required per person per day and that 5 g per day is potentially available per person. In West Africa where the population is smaller and production is very high, a greater proportion of the requirement would be satisfied, and it has been estimated, for example, that about 10 g per day is potentially available per person in Nigeria (Milner, 1964). However, groundnut is an important cash-crop in these countries and much of the production is in fact exported. The expanded cultivation in the United States subsequent to the Civil War, followed by the introduction of groundnut to the European markets in 1840, has led to the development of large-scale industries based on oil extraction. The principal groundnut producing countries are listed in Table 11.1.

The use of the by-products from the oil-expelling industry as animal food led to extensive work on the nutritional evaluation of the material not only as a protein food in its own right but also as an adjunct to other feedstuffs such as cereals; especially for pigs and poultry. Whole groundnuts are commonly used as a feed for pigs in the United States, but the large oil content results in the production of soft pork. The composition of groundnut kernels is given in Table 11.2. The extracted meal is a more suitable feed although the amino acid composition of groundnut protein compares unfavourably with most other plant protein sources as a supplier of the essential amino acids required by

Table 11.1. *Groundnut production in principal producing countries (10³ tonnes)*

	1967	1968	1969	1970	1971
India	5731	4631	5130	6065	5800†
China	2300†	2150†	2350†	2650†	2700†
Nigeria	1256†	1445†	1365†	780†	1100†
USA	1122	1153	1147	1351	1357
Senegal	1005	830	796	583	960
Brazil	751	754	754	928	850‡
Burma	389	398	444	520	520‡
Indonesia	402	478	445	488	480‡
S. Africa	429	227	368	318	399
Argentina	354	283	217	235	388
Sudan	297	197	383	353†	353‡
Uganda	200†	234	234†	234†	234‡

† Unofficial figure. ‡ FAO estimate.
After Production Yearbook (1971) No. 25, p. 232. FAO Rome.

Table 11.2. *Composition of groundnut kernels* (%)

	Range	Mean
Moisture	3.9–13.2	5.0
Protein	21.0–36.4	28.5
Lipids	35.8–54.2	47.5
Crude fibre	1.2–4.3	2.8
Nitrogen-free extract	6.0–24.9	13.3
Ash	1.8–3.1	2.9
Reducing sugars	0.1–0.3	0.2
Disaccharides	1.9–5.2	4.5
Starch	1.0–5.3	4.0
Pentosans	2.2–2.7	2.5

After Freeman, A. F., Morris, N. J. & Willich, R. K. (1954). *Peanut Butter*. USDA Paper No. AIC-370. Southern Regional Research Laboratory, New Orleans, Louisiana.

non-ruminant animals (Table 11.3). Growth depressing substances such as a trypsin-inhibitor are present (Borchers & Ackerson, 1950), but it seems likely that normal processing of groundnut meal adequately inactivates the inhibitor (Woodham & Dawson, 1968).

Many evaluations of the nutritional quality of groundnut meal have been made using rats, chickens and pigs. These have included comparisons of a number of samples of groundnut meal with each other and with other types of protein concentrates (Duckworth, Woodham & McDonald, 1961), estimations of the limiting amino acids (Black & Cuthbertson, 1963; Ellinger & Boyne, 1963; Milner & Carpenter, 1963;

Table 11.3. *Ranges for levels of amino acids, essential for the non-ruminant, in extracted groundnut meal compared with soybean, cotton-seed and sunflower seed meals (g/16 g N)*

	Groundnut	Soybean	Cottonseed	Sunflower
No. of samples ...	(8)	(7)	(6)	(2)
Threonine	2.6–3.5	3.8–4.3	3.1–3.7	3.6–3.8
Glycine	5.2–6.5	3.9–4.7	4.3–5.6	5.5–5.7
Valine	3.6–4.9	4.7–5.5	4.8–5.2	4.6–5.0
Cystine	1.1–1.4	1.3–1.6 (5)	1.9 (1)	1.9–2.2
Methionine	0.7–1.0	0.9–1.5	1.0–1.7	2.4 (2)
Isoleucine	3.4–4.3	4.2–5.2	3.4–3.9	3.9–4.3
Leucine	5.9–6.8	7.4–8.4	6.3–7.3	6.2 (2)
Tyrosine	3.9–5.6	3.4–4.0	2.4–4.0	2.5–2.7
Phenylalanine	4.3–7.3	5.1–6.0	5.6–6.8	4.3–4.8
Lysine	3.2–4.0	6.0–6.9	3.8–5.2	3.5–3.7
Histidine	2.1–2.7	2.1–3.4	2.7–3.3	2.4–2.6
Arginine	10.0–16.0	7.1–8.5	10.8–13.0	8.5–8.7
Tryptophan	1.1 (1)	1.9 (1)	—	1.8 (1)

Ranges include all samples except where the actual number analysed is indicated in brackets. All analyses were carried out at the Rowett Research Institute using a Technicon NC1 Auto-analyser for all amino acids excepting tryptophan. Cystine and methionine were estimated on samples oxidised with performic acid. Tryptophan was estimated colorimetrically by a modification of the procedure of Spies & Chambers (1949).

Anderson & Warnick, 1965; Wessels, 1967) and studies of amino acid availability (Carpenter, McDonald & Miller, 1972).

All of these studies agree in defining groundnut as a generally poor protein source for non-ruminants compared with other plant protein foods such as soybean and cottonseed. This is chiefly ascribed to low levels of lysine and of the sulphur amino acids – the essential amino acids most likely to be lacking in animal rations based upon cereals.

A number of studies have been made, some forming part of the UK contribution to IBP, into the effects of environment on groundnut proteins as well as into the differences between groundnut varieties. The amount of protein in varieties grown in India varies from 43.8 to 65.3 % but no significant differences were found in biological value as predicted from the methionine content (Cheema & Ranhotra, 1967). Other Indian workers have also reported that varietal differences in amino acid composition are small and rat feeding trials confirmed this conclusion (Chopra & Sidhu, 1967).

The chief proteins in groundnut are contained in fractions which are called arachin and conarachin, the former predominating quantitatively (Tombs, 1965; Dawson, 1969). Significant differences in the composition of the fractions have been demonstrated depending upon variety and location of growth. The same variety grown in different areas may yield, for instance, conarachin fractions exhibiting different electrophoretic patterns (Dawson & McIntosh, 1973). The chemical scores of such 'conarachins' calculated from amino acid analysis ranged from 68 to 82; this suggests that there is some scope for selecting groundnut varieties for growth in a particular locality. The IBP has been instrumental in furthering the establishment of a gene pool at Samaru in Nigeria by encouraging the collection, screening and maintenance of varieties and cultivars from various sources. The importation of the new varieties has already resulted in the establishing of varieties capable of yielding 1680–2240 kg/ha. Further work is required in order to see if such high-yielding varieties are also of superior nutritional quality.

Processing of groundnuts

Traditionally most oilseeds were processed primarily for their oil, the residual cake or meal being sold either as animal feed or, if heat damaged or grossly infected, as manure. The vast potential of oilseed meals, if suitably upgraded, in fighting human protein malnutrition has been well recognised (e.g., Anson, 1962). This recognition, together with increasing demands for cheap, food-grade protein additives in the manufacture of 'convenience' foods, has led to some refinement in the conventional standard oil-milling techniques. Thus a small but increasing proportion of the total oilseed meal production is now of a quality suitable for inclusion in human foods.

Conventional processing

In the conventional industrial crushing of groundnuts for oil and cake (e.g., Rosen, 1958), nuts are screened to remove stones, trash and metal, then shelled and steam-cooked to coagulate the protein and free the oil. Oil extraction is either by mechanical expression, usually in a continuous screw press (expeller), or by solvent extraction, usually done after preliminary low-pressure expelling (pre-press solvent extraction). Satisfactory solvents include those of the hexane- and cyclohexane-types: trichloro-ethylene gives a dark oil containing more phosphatide, a valuable by-product, but the meal so produced may be toxic. Solvent is stripped from the meal by steam-heating.

Table 11.4. *Groundnut cake, meal, flour and protein isolate compositions*

	*Decorticated expeller cake (%)	*Decorticated extracted meal (%)	†High quality flour from USA nuts (%)	‡Protein isolate (%)
Moisture	10	10	8	3.4
Crude protein (N × 6.25)	45.4	49.7	60	95
Oil	6.0	0.7	0.75	0.5
Crude fibre	6.5	7.9	4.5	—
Nitrogen free extractives	26.4	26.0	22.5	—
Ash	5.7	5.7	4.5	0.5

* Evans (1960). † Woodroof (1966). ‡ De & Cornelius (1971).

Food-grade processing

In food-grade processing, additional operations and modifications to existing equipment are required to ensure a sanitary product which retains an optimal protein nutritive value, is free from toxic factors and which at the same time possesses acceptable physical and organoleptic properties.

Food-grade groundnut protein products comprise flours (45–60 % CP), isolates (90–95 % CP) and the lipid–protein isolate 'Lypro' (65 % CP). The composition of these products is given in Table 11.4. Having the lowest production costs and retaining the insoluble carbohydrate fraction, the flours must be regarded as the most useful of these products in emergency feeding programmes, where the supplementary protein is to be incorporated into balanced mixed foods (Anson, 1662). Protein and lipid–protein isolates, because of their higher production costs, and because their physical properties can be more readily tailored to needs, are more appropriately used in the preparation of beverages – toned or artificial milks – or of ice-creams. They also have wide non-nutritional uses in food technology, by virtue of their functional properties.

Existing oil-milling plants may usually be adapted to produce food-grade groundnut flour. In practice this means: (1) extra processing to give clean, white 'meats' for pressing; (2) overall improvements in plant hygiene; (3) improvements in plant control. Thus extra control and care is needed during cooking and pressing to ensure that the maximum temperature does not exceed 120 °C; and similar control is also needed during solvent extraction, which must be done with hexane of optimum quality (PAG, 1971). Examples of specifications and costings for up-grading existing mills are given by De & Cornelius (1971), whose excellent review provides *inter alia* specifications and costings for new plants and breakdowns of processing costs in different countries. Woodham & Dawson (1968) provide a brief summary of processing temperatures in commercial plants, with regard to the nutritive value of groundnut protein.

Groundnut flour

Before purchasing, groundnuts intended for food-grade flour production must be adequately sampled and shown to be essentially free from mould damage and from pest-infestation. In the factory, storage must be under cool, dry conditions, avoiding mould, insect and rodent attack. Kernels separated from cleaned, decorticated nuts are blanched (generally by heating), to remove testas and germs which contain anti-growth principles and bitter flavour principles, respectively. Discoloured 'meats' must now be removed, and colour-sensing electronic separators are available, which are said to be efficient in rejecting aflatoxin infected nuts (Pattinson *et al.*, 1968). A similar colour-sorting step may also be needed before blanching. The white 'meats' are now ready for cooking and pressing, as outlined previously.

Groundnut protein isolates

In conventional processes (De & Cornelius, 1971) food-grade flour or cake is extracted with dilute NaOH solution and the insoluble fibrous carbohydrate fraction separated by centrifugation or filtration. The groundnut proteins are precipitated from solution either by lowering the pH of the filtrate to a value of about 4.5 or, adopting the classic Chinese procedure, by heating in the presence of Ca^{2+}. Advantages of the latter procedure are that it may assist the elimination from the isolate of 'beany' flavours and aflatoxin; it may also help to give a protein with 'chewy gel' texture which in turn may facilitate the production of fibrous meat analogues (Anson, 1962).

The Chayen process (Smith, 1966), uses as feedstock, blanched meats which are disintegrated in dilute NaOH solution by impulse rendering. The liquor from the impulse renderer is separated by high-speed centrifugation into three streams carrying carbohydrate, oil emulsion and lipid–protein complex. This is precipitated in line, thickened by centrifugation, neutralised, sterilised and finally spray-dried to give the product 'Lypro' (65% CP; 32% lipid). A commercial plant producing Lypro, with a throughput of 1 tonne of raw kernels per hour was operated by International Protein Products at Plymouth between 1960 and 1967. In 1966 Lypro sold at about £280 per tonne. The minimum size of economic unit for the Chayen process is estimated to be a plant with an output of 8300 tonnes of Lypro per year. Solvent extraction may be used to convert Lypro into a conventional isolate (95 % CP).

The CFTRI process (Orr & Adair, 1967; De & Cornelius, 1971), essentially based on the Chayen process, claims to produce an isolate of higher protein content (92 %) with better recovery of the free oil (91–4 %). A plant designed to produce 2 tonnes of isolate per day was operated by the Tata Oil Mill Company of Bombay, but did not realise these claims. A modified plant at Ahmedabad now produces 100 kg of isolate per day but the oil recovery is only 80–2 %.

Groundnut flours and isolates in protein food programmes

De & Cornelius (1971) discuss this topic with special reference to Nigeria, Senegal and India. Other reviews include those by Chandrasekhara & Ramanna (1969) and by Orr & Adair (1967) who, like De & Cornelius, give tentative specifications for the quality of groundnut flour for human food. Those are summarised in Table 11.5. It is unfortunately only too evident that the pathogenic mould, *Aspergillus flavus*, imposes severe constraints on the ready exploitation of the groundnut as a cheap protein foodstuff.

Aflatoxin

'Turkey-x' disease, an apparently new disease of obscure aetiology, resulted in 1960 in the disastrous loss of 100000 Norfolk turkey poults in England. A toxic Brazilian groundnut meal was later established as the cause of this disease; and subsequently the agent was found to be a group of fungal metabolites – aflatoxin – produced by an infecting mould, *Aspergillus flavus*. Four closely-related compounds, aflatoxins B_1, B_2, G_1 and G_2, are commonly associated with groundnuts thus infected. Aflatoxins M_1 (4-hydroxy B_1), with similar toxicity to B_1, and M_2 (4-hydroxy B_2), are produced by ruminant animals from ingested B_1 and B_2 and are excreted in the milk (Fig. 11.1).

The aflatoxins are primarily liver poisons and are acutely toxic to most species, although there is a wide range of susceptibilities between different species. The comparative hepatotoxic effects of aflatoxin in farm animals have been summarised by Allcroft & Lancaster (1966). Allcroft (1969) also tabulates data on the effect on farm animals of relatively low dietary levels of aflatoxin. These data should facilitate

Table 11.5. *Quality specifications and control for food-grade groundnut flour*†

Properties

(*a*) Colour:	light tan or buff
(*b*) Odour:	free from mustiness, solvent or unpleasantness
(*c*) Taste:	should be smooth on tongue
(*d*) Particles:	no coarse or gritty particles should be present

Percentage composition

Moisture (range) 7.00–11.0
Crude fat (wax) 8.0
Crude protein (N × 6.25) (min) 48.0
Crude fibre (max) 3.5
Ash (max) 4.5
Free fatty acids (max) 1.0 (based on oil)

Protein quality

(*a*) Available lysine
(*b*) Protein dispersibility index

Processing to be carried out to maximise these values; minimum permissible figures have not been specified.

Aflatoxin

Maximum permissible limit of 30 μg/kg.

Bacteriology

Product should be free of *E. coli*, *Salmonella* and pathogens. Total bacterial count of not more than 20000 organisms/g.

Acid-insoluble ash

Not to exceed 1 % of sample weight.

Insect and rodent contamination

Product to be essentially free of insects, insect fragments, rodent hairs and excreta.

† Based on PAG (1961), after De & Cornelius (1971).

the estimation of tolerated inclusion levels of toxic groundnut meal in feeds for different classes of livestock.

In certain species, including ducklings, trout and rats, aflatoxins act as potent carcinogens; thus as little as 0.015 ppm in the diet led, in chronic feeding experiments, to 100 % incidence of liver carcinoma in rats. In addition rats fed on a diet of low protein or lipotrope content have been found to be more susceptible to aflatoxin than controls receiving normal diets (Madhavan & Gopalan, 1965; Newberne & Rogers, 1970), a finding which has possible bearing on the epidemiology of primary liver carcinoma in humans. Thus although there is at present no direct evidence that aflatoxin is carcinogenic in man, there is a great incidence of primary liver cancer in the native populations of tropical areas where malnutrition may be combined with a dietary intake of aflatoxin, and studies are now being made of this possible relationship between the disease, diet and toxin ingestion.

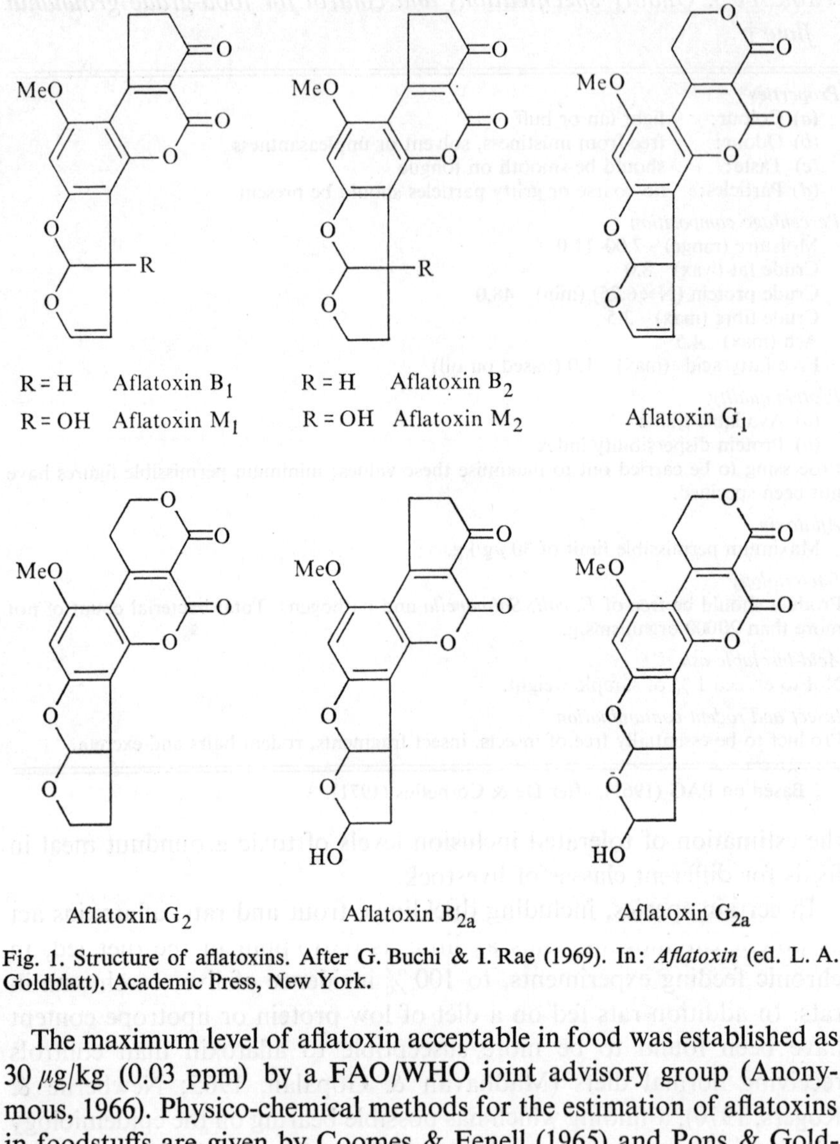

R = H Aflatoxin B_1
R = OH Aflatoxin M_1

R = H Aflatoxin B_2
R = OH Aflatoxin M_2

Aflatoxin G_1

Aflatoxin G_2

Aflatoxin B_{2a}

Aflatoxin G_{2a}

Fig. 1. Structure of aflatoxins. After G. Buchi & I. Rae (1969). In: *Aflatoxin* (ed. L. A. Goldblatt). Academic Press, New York.

The maximum level of aflatoxin acceptable in food was established as 30 μg/kg (0.03 ppm) by a FAO/WHO joint advisory group (Anonymous, 1966). Physico-chemical methods for the estimation of aflatoxins in foodstuffs are given by Coomes & Fenell (1965) and Pons & Goldblatt (1969); Legator (1969) reviews the biological assay for aflatoxins.

Control of aflatoxin

Moulds of the *Aspergillus flavus* group (mainly *A. flavus* Link & *A. parasiticus*) grow and produce aflatoxins in a wide range of crops, including grains, pulses, food staples and oilseeds, but especially groundnuts and cottonseed (Diener & Davis, 1969; Golumbic & Kulik, 1969). The main factors leading to high contamination levels in groundnuts are shell damage and kernel splitting, usually induced by insects, poor harvesting or drought. Because *A. flavus* will grow only on groundnuts containing more than 9 % moisture, aflatoxin contamination can be minimised by prompt harvesting and drying and careful storage of the nuts under dry, cool conditions. In practice traditional ideas on groundnut handling may conflict with this prophylactic approach, so that, subsequently, less economic methods involving sorting (Kensler & Natoli, 1969), or detoxification (Dollear, 1969), must be adopted to reduce the risk of aflatoxicosis.

References

Allcroft, R. (1969). In: *Aflatoxin* (ed. L. A. Goldblatt), p. 237. Academic Press, New York.

Allcroft, R. & Lancaster, M. C. (1966). *International Microfilm Journal of Legal Medicine*, **1**, Item 12.

Anderson, J. O. & Warnick, R. E. (1965). *Poultry Science*, **44**, 1066.

Anonymous (1966). *Nature, London*, **212**, 1512.

Anson, M. L. (1962). *Archives of Biochemistry & Biophysics*, Supplement **1**, 68.

Black, A. E. & Cuthbertson, W. F. J. (1963). *Proceedings of the Nutrition Society*, **22**, xxi.

Borchers, R. & Ackerson, C. W. (1950). *Journal of Nutrition*, **41**, 339.

Carpenter, K. J., McDonald, I. & Miller, W. S. (1972). *British Journal of Nutrition*, **27**, 7.

Chandrasekhara, M. R. & Ramanna, B. R. (1969). *Overdrutuit Voeding*, **30**, 297.

Cheema, P. S. & Ranhotra, G. S. (1967). *Journal of Nutrition and Dietectics*, **4**, 93.

Chopra, A. K. & Sidhu, G. S. (1967). *British Journal of Nutrition*, **21**, 519, 583.

Coomes, T. J. & Fenell, A. J. (1965). TPI Report No. G13. Tropical Products Institute, London.

Dawson, R. (1969). *Proceedings of the Nutrition Society*, **28**, 4A.

Dawson, R. & McIntosh, A. D. (1973). *Journal of the Science of Food and Agriculture*, **24**, 597.

De, S. S. & Cornelius, J. A. (1971). Agricultural Services Bulletin No. 10, FAO, Rome.

Diener, U. L. & Davis, N. D. (1969). In: *Aflatoxin* (ed. L. A. Goldblatt), p. 13. Academic Press, New York.

Dollear, F. G. (1969). In: *Aflatoxin* (ed. L. A. Goldblatt), p. 359. Academic Press, New York.

Duckworth, J., Woodham, A. A. & McDonald, I. (1961). *Journal of the Science of Food and Agriculture*, **12**, 407.
Ellinger, G. M. & Boyne, E. B. (1963). *Proceedings of the Nutrition Society*. **22**, xiii.
Golumbic, C. & Kulik, M. M. (1969). In: *Aflatoxin* (ed. L. A. Goldblatt), p. 307. Academic Press, New York.
Greenwood, M. (1949). Commonwealth Bureau of Soil Science Technical Communication No. 46.
Greenwood, M. (1951). *Empire Journal of Experimental Agriculture*, **19**, 225.
Kensler, C. J. & Natoli, D. J. (1969). In: *Aflatoxin* (ed. L. A. Goldblatt). p. 333. Academic Press, New York.
Legator, M. S. (1969). In: *Aflatoxin* (ed. L. A. Goldblatt), p. 107. Academic Press, New York.
Madhavan, T. V. & Gopalan, C. (1965). *Archives of Pathology*, **80**, 123.
Meredith, R. M. (1964). *Empire Journal of Experimental Agriculture*, **32**, 126.
Milner, C. K. & Carpenter, K. J. (1963). *Proceedings of the Nutrition Society*, **22**, xii.
Milner, M. (1964). *Protein enriched foods for world needs*. American Association of Cereal Chemists, St Paul, Minnesota.
Newberne, P. M. & Rogers, A. E. (1970). In: *Mycotoxins in human health* (ed. I. F. H. Purchase). Council for Scientific & Industrial Research, Pretoria.
Orr, E. & Adair, D. (1967). TPI Report G31. Tropical Products Institute, London.
Oyenugu, V. A. (1967). *Agriculture in Nigeria*. FAO, Rome.
Oyenugu, V. A. (1968). *Nigeria's food and feedingstuffs: their chemical and nutritional values*, 3rd edition. University Press, Ibadan.
PAG (1961). *Guideline for processing and quality of edible groundnut flora*. Protein Advisory Group of the United Nations, New York.
PAG (1971). Document 3. 14/14 Appendix G, p. 79. Protein Advisory Group of the United Nations, New York.
Pattinson, I., Crowther, P. & Noubey, H. (1968). *Tropical Science*, **10**, 212.
Pons, W. A. & Goldblatt, L. A. (1969). In: *Aflatoxin* (ed. L. A. Goldblatt), p. 77. Academic Press, New York.
Rosen, G. D. (1958). In: *Processed plant foodstuffs* (ed. A. M. Altschul). Academic Press, New York.
Smith, R. H. (1966). Advances in Chemistry Series No. 57, p. 133. American Chemical Society, Washington.
Spies, J. R. & Chambers, D. C. (1949). *Analytical Chemistry*, **21**, 1249.
Tombs, M. P. (1965). *Biochemical Journal*, **96**, 119.
Wessels, J. P. H. (1967). *South African Journal of Agricultural Science*, **10**, 113.
Woodham, A. A. & Dawson, R. (1968). *British Journal of Nutrition*, **22**, 589.

12. Broad bean

A. HAGBERG & J. SJÖDIN

Broad bean is grown in many parts of the Old World (Europe and Asia) as well as in North America. In 1956, for instance, China produced 3 Mt, Italy 285000 t and Egypt 208000 t (Deschamps, 1958). In Scandinavia broad beans occupy about 15000 ha. The small-seeded type (*Vicia faba* L. var. minor Beck) is the main type grown. In Scandinavian conditions it yields 3000 to 4000 kg/ha. In terms of both protein and dry matter this is more than the yield from any other big-seeded legume grown in Europe.

The content of crude protein (N × 6.25), as percentage of dry matter, is 28 to 33 % (Bingefors & Sjödin, 1969). The average figure in Sweden is 30.0 % for the standard variety Primus. Thus the protein yield under Scandinavian conditions is 900 to 1200 kg/ha. Broad beans contain around 57 % carbohydrates (starch about 42 %), 2 % oil, 8 % crude fibre and about 4 % ash (White, 1966). The crude fibre is mainly concentrated in the seed-coat, constituting about 14 % of the seed. The seed-coat is poor in protein (Clausen & Hansen, 1968).

There is metabolic and storage protein in legume seeds (Millerd, 1972). The latter is mainly globulin, insoluble in water and soluble in dilute salt solution, situated in the cotyledons. In both *Vicia* and *Pisum* the globulins are distinguished as legumin with molecular weight about 350000 and vicilin with molecular weight about 180000 in the ratio of 2:1 (Danielsson, 1949). Much of the protein can be extracted and is processed into 'textured vegetable protein' for use as human food. Among the important amino acids in broad bean the following values, expressed as percentages, may be mentioned (Eggum, 1968): lysine, 6.3; methionine, 0.9; cystine, 1.5; threonine, 3.6; and tryptophan, 0.9. Compared with cereals, e.g., barley, the main difference is that there is more lysine in *Vicia faba*, and more methionine in barley. The biological value (the percentage of the protein that can be used by growing rats), is 50 (Eggum, 1968), under conditions in which the value for barley was 72, for soybean meal 62 and for egg 99. This is a consequence of the deficiency in sulphur-containing amino acids.

In feeding experiments on mice (Löfqvist & Munck, 1969) protein concentrates of broad bean and soybean, supplemented with 1 % methionine, were compared with a casein standard. Mice fed on soybean showed a somewhat lower weight gain and a somewhat higher quantity of consumed protein compared to casein, whereas the mice fed on broad bean showed a higher weight gain and a higher protein consumption compared to casein. In both cases the PER (protein efficiency ratio) was somewhat lower for the legumes. A comparison between concentrates and raw material of these two legumes showed no difference for broad bean irrespective of methionine supplementation, whereas the processing increased the nutritional value for soybean.

The last experiments suggest that there probably are more growth inhibiting substances in soybean than in broad bean. Feeding trials with broad beans on other animals have shown that there are upper limits to the content of broad beans in the diets for normal weight gain. The reason for this is probably the presence of 'toxic' or inhibiting substances. It is supposed that these are mainly haemagglutinins and antitrypsins. Tannins and cyanogenic glucosides have also been discussed (Becher & Nehring, 1969). With dairy cows (Sörenssen & Kristensen, 1968) 30 % broad beans in the feed mixture can be used without any negative influence on milk production and fat content. The low fat content in the broad bean has then to be compensated for in the fodder. Broad beans can also be used for young stock (Lykkeaa & Sörenssen, 1968). For pigs (Hansen, 1967) 15 % broad beans in the diet seems to be an upper limit and this amount can also be used for poultry if methionine is added (Mahon & Common, 1950; Blair & Bolton, 1968). Preparations that are to be used as human food in parts of the world where the genetic trait 'favism' is common will need careful screening to ensure that they do not carry the factor(s) responsible for toxicity.

Plant breeding projects have so far been concerned with early ripening and type of growth to make combine harvesting easy in North and West European conditions. The great problem from a nutritional point of view is to increase the methionine in broad bean protein and thus improve the biological value. So far we have had no efficient screening technique for methionine, but this problem now seems to be solved. Thus, it should be possible to start intensive selection.

References

Becher, M. & Nehring, K. (1969). *Handbuch der Futtermittel* 2. Verlag Paul Parey, Hamburg.

Bingefors, S. & Sjödin, J. (1969). *Journal of the Swedish Seed Association*, **3–4**, 218–30.

Blair, R. & Bolton, W. (1968). *Journal of Agricultural Science*, **71**, 355–8.

Clausen, H. & Hansen, V. (1968). *Landökonomisk forsöglaboratorium. Årsbog*, pp. 45–58. Landökonomisk forsögslaboratorium, Copenhagen.

Danielsson, C. E. (1949). *Biochemical Journal*, **44**, 387–99.

Deschamps, I. (1958). In: *Processed plant protein foodstuffs* (ed. A. M. Altschul), pp. 717–37. Academic Press, New York.

Eggum, B. O. (1968). *Aminosyrekoncentration og proteinkvalitet*. Stougaards forlag, Copenhagen.

Hansen, V. (1967). *Landökonomisk forsögslaboratorium. Årsbog*, pp. 85–91. Landökonomisk forsögslaboratorium, Copenhagen.

Löfqvist, B. & Munck, L. (1969). *Journal of the Swedish Seed Association*, **3–4**, 231–47.

Lykkeaa, J. & Sörenssen, M. (1968). *Landökonomisk forsögslaboratorium. Årsbog*, pp. 585–7. Landökonomisk forsögslaboratorium, Copenhagen.

Mahon, J. H. & Common, R. H. (1950). *Scientific Agriculture*, **30**, 43–9.

Millerd, A. (1972). In: Symposium on the biology of plant storage proteins, Canberra, 1972.

Sörenssen, M. & Kristensen, V. F. (1968). *Landökonomisk forsögslaboratorium. Årsbog*, pp. 490–503. Landökonomisk forsögslaboratorium, Copenhagen.

White, H. L. (1966). *Journal of Experimental Botany*, **17**, 195–203.

13. Concentrates by wet and dry processing of cereals*

R. M. SAUNDERS & G. O. KOHLER

The search for protein sufficiency has stimulated investigation of many unusual sources of protein, but relatively few investigators have focused their attention on the most obvious and readily available source of all, cereal grains. Cereal grains represent at present the most plentiful and most economical source of protein suitable for man's needs. Table 13.1 shows the world production of cereal grains in 1971.

Corn

Of the annual production of corn (*Zea mays*) in the United States, about 5 % is milled wet, and 3.5 % dry and, in part, converted into high protein feedstuffs. These large-scale commercial processes have been described in detail by Inglett (1970). Table 13.2 shows the nutrient analysis for different products obtainable. Condensed fermented corn extractives or heavy corn steepwater are obtained by partial dehydration of the liquid that results from steeping corn in a water and sulfur dioxide solution that is allowed to ferment by the action of naturally occurring lactic-acid-producing micro-organisms (as practiced in the wet milling of corn). Corn gluten feed is that part of the corn that remains after the extraction of the larger portion of the starch, gluten and germ by the processes employed in the wet milling manufacture of starch. Corn gluten meal is the residue from corn after removal of the larger part of starch and germ, and the separation of the bran by the process employed in the wet milling manufacture of starch, or by enzymatic treatment of endosperm. Corn germ meal (wet milled) is ground corn germ from which the solubles have been removed by steeping out most of the oil (removed by hydraulic, expeller or solvent extraction processes), and is obtained in the wet milling process of manufacture of starch, syrup or other corn products. Corn gluten meal also has been defatted and used as a high protein ingredient for humans. Two products, containing 70–5 % protein and 90–5 % protein have been manufactured (Reiners *et al.*, 1972). Corn distillers dried grains with solubles are the product obtained after the removal of ethyl alcohol

* Literature reviewed through 1972.

Table 13.1. *World production of cereal grains in 1971*

Commodity	Production (Mt)
Corn (*Zea mays*)	291
Barley (*Hordeum vulgare*)	130
Oats (*Avena sativa*)	55
Sorghum (*Sorghum vulgare*)	42
Wheat (*Triticum aestivum*)	322
Rice (*Oryza sativa*)	289
Rye (*Secale cereale*)	29

Table 13.2. *Nutrient analysis of protein concentrates from corn*†

Material	% Protein (N × 6.25)	% fat	% fiber	% ash	% starch
Condensed fermented corn extractives	47.2	0	0	14.7	—
Corn gluten feed	24.8	2.8	8.9	8.1	15.6
Corn gluten meal; Process 1	46.9	2.8	5.1	4.4	—
Corn gluten meal; Process 2	68.9	2.8	1.3	2.0	—
Corn germ meal	25.1	2.1	10.5	4.2	26.7
Corn distillers dried grains with solubles	29.7	8.8	9.3	4.9	—

† Dry basis. Shroder & Heiman (1970).

by distillation from the yeast fermentation of corn or a corn–grain mixture.

Corn protein is deficient in lysine and tryptophan, but supplementation with soybean protein is an accepted practice. Although not a corn protein concentrate *per se*, a blend of corn, soy and dried skim milk, known as CSM, is widely available throughout the world as a highly nutritious foodstuff, with a PER equal to that of casein (Inglett, 1970).

Rice

The commercial processes for milling rice (*Oryza sativa*) are described in a recent monograph by Houston & Kohler (1970). High protein flours have been prepared with two types of commercial machinery from different rices including high protein rice. One process, consisting of fine-grinding and air classification, furnished about 10 % of the kernels as flour with higher protein than normal rice flour. The second process involves scouring off consecutive layers of milled rice in a continuous-flow abrasive cone mill similar to a rice-whitening cone.

Table 13.3 *Yield and protein content of flours obtained by special milling of rice*†

Milling procedure	Flour % yield	Flour % protein (N × 5.95)
Air-classified Pearl rice (5.8 % protein)	11.9	8.4
Abraded rice (6.0 % protein)	12.5	12.3
Abraded rice (9.6 % protein)	12.8	16.2
Abraded, then air-classified rice (9.6 % protein)	4.3	18.9

† Dry basis. Houston *et al.* (1964).

Table 13.4. *Yield and protein content of flours obtained by special milling of rice bran and polish*†

Milling procedure	Flour % yield	Flour % protein (N × 5.95)
Air classification of rice polish (10.4 % protein)	59.8	10.8
Air classification of defatted bran (10.6 % protein)	47.8	12.3
Sieve classification of defatted bran (10.6 % protein) through a 40-mesh screen	52.5	13.4

† Dry basis. Houston & Mohammad (1966).

Flour from the outermost layers has about twice the original protein content of the whole kernels, and is obtained in 10–15 % yield. Fine-grinding and air classification of flours obtained by abrasive milling can further concentrate protein in the finest-particle fractions. Fractions containing increased protein levels also have higher levels of fat, ash and vitamin B. Table 13.3 illustrates the yield and protein content of products obtainable by these processes. The flours obtained by abrasive milling can be produced by available commercial equipment from any type of milled rice. This includes long-, medium-, and short-grain regular white rice, waxy rice, and parboiled rice. The abrasion process appears to be more attractive than air classification of finely ground rice. It has been suggested they would be suitable for infant and geriatric foods.

Air and sieve classification of defatted and finely ground rice bran or polish yields products enriched in protein (Table 13.4). It seems likely that the sieving process could be advantageous in regions in need of quality protein.

Concentrates by mechanical extraction

Protein concentrates have been prepared from rice bran and defatted bran by a wet process (Chen & Houston, 1970; Lew *et al.*, 1971; Saunders *et al.*, 1972). The bran is extracted in an alkaline medium and fractionated by a combination of isoelectric precipitation (pH 5.5) and heat coagulation. In defatted bran, the percentage of protein in the total extracted solids increased from 28.9 % at pH 7.0 to 45.5 % at pH 11. Extraction at pH 11 followed by neutralization and drying of the total extract recovered from a defatted bran with 10 % protein about 50 % of the total protein as a 40 % protein concentrate. By isoelectric precipitation, an 85 % protein concentrate can be obtained, corresponding to 37 % of the original protein (Chen & Houston, 1970). Extraction of full-fat rice bran with alkali can remove up to 71 % of the bran protein as pH is increased from 7 to 12. At pH 11, the total extractable solids, after drying, contain 58 % of the bran protein as a 35 % protein concentrate. Isoelectric precipitation at pH 4.5 yields a 62 % protein concentrate containing 44 % of the original bran protein. Washing the latter product with alcohol yields a light tan product containing, on a dry basis, 76 % protein, 14 % lipid and 6 % carbohydrate, and negligible fiber (Lew *et al.*, 1971). This simple extraction procedure could make available considerable amounts of good quality food protein in rice-growing regions where rice bran is not generally defatted.

Oats

Protein concentrates have been made from oats (*Avena* spp.) by both wet and dry processes (Wu *et al.*, 1973*a*, *b*; Cluskey *et al.*, 1973). Both oat groats and first and second flours, from a high protein variety and a normal variety, yield protein-enriched fractions after being finely ground and air classified (Table 13.5). All the protein concentrates contain adequate amounts of sulfur amino acids and lysine.

A wet milling process yields a protein concentrate from dry milled groats, both full-fat and defatted (Wu *et al.*, 1972). Hammer milled groats are first extracted with dilute alkali at pH 9. The juice is filtered to remove the branny residue and the resultant juice is centrifuged to remove the starch. The supernatant at this stage contains dissolved protein; it is adjusted to pH 6 and taken to dryness. Table 13.6 lists the yields and protein content of products obtainable by this simple procedure. The good quality of the protein in oats is an added bonus for these concentrates.

Table 13.5. *Yield and protein content of concentrates prepared by air classification of selected oat fractions*†

Sample	% yield	% protein (N × 6.25)	% yield based on whole groats
Normal variety			
Ground groats (16.3 % protein)	27	24.3	27
First flour (10.1 % protein)[a]	27	19.6	12
Second flour (15.9 % protein)[a]	29	29.7	7
High-protein variety			
Ground groats (22.7 % protein)	28	32.8	28
First flour (15.8 % protein)[b]	30	31.8	15
Second flour (28.9 % protein)[b]	34	45.6	7

† Dry basis. [a] Represents 43 % (first) and 25 % (second) of ground groats. [b] Represents 51 % (first) and 20 % (second) of ground groats.

Table 13.6. *Protein concentrates obtained by alkaline extraction of full-fat and defatted oat groats*

Sample	% yield†	% protein (N × 6.25)	% original protein
Full-fat groats from moderate protein variety	21	59.1	66.2
Defatted groats from high protein variety	19	75.0	70.0

† Dry basis.

Wheat

One of the most important developments in flour milling has been the use of air classification to manufacture from an ordinary wheat (*Triticum aestivum*), flours which differ in protein content and accordingly suit different purposes. Particle size is the predominant factor, but yield and protein content of the fractions are affected by the type of wheat. Table 13.7 summarizes data showing the concentration of protein in certain fractions (Ziegler & Greer, 1972). Concentrates of this type are made commercially. It must be emphasized, however, that factors other than protein are shifted to the fine fraction, notably damaged starch, discoloring matter, fat and vitamins (Ziegler & Greer, 1972). These affect the value of a fraction for a particular purpose.

Kent & Evers (1969) have also shown that by differences in particle density it is possible to separate from the coarse fractions of air classified

Table 13.7. *Yield and protein content of fractions obtainable by air classification of wheat flours with and without pinned-disc grinding*†

Flour	Fractions	Particle size	% yield	% protein (N × 5.7)
Soft wheat (9.6 % protein)	Unground flour	Fine	9	18.5
		Medium	34	4.5
		Coarse	57	11.3
	Ground flour	Fine	20	21.9
		Medium	68	6.6
		Coarse	12	10.3
Hard wheat (13.1 % protein)	Unground flour	Fine	3	20.2
		Medium	14	11.4
		Coarse	84	13.5
	Ground flour	Fine	14	23.9
		Medium	51	10.0
		Coarse	35	11.6

† Dry basis.

flours, material of high protein content. Although the yield is very small, specimens having a protein content greater than 50 % can be obtained.

A wet process to separate protein from starch in wheat flour on a continuous basis has been described (Johnston & Fellers, 1971). Essentially a modification of the Fesca process, flour is mixed with an optimum amount of water then finely dispersed by shearing action. The starch is centrifuged out, and the solubles dried. The protein concentrate thus obtained contains from 20 to 40 % protein depending on the flour composition. Functional and nutritional characteristics of this protein concentrate make it a desirable food ingredient. The other product of this process, a high quality starch, should find ready acceptance in conventional foods.

In the production of regular white flours most of the high quality protein in germ and aleurone goes into the by-products customarily used for animal feeds. Approximately one million tonnes of protein is expended this way. Fellers *et al.* (1966*a*, 1968) have described a process whereby through further milling and sifting, wheat millfeeds† can provide flours high in protein, low in fiber, and suitable for use in food products. The yield and composition of the flours varies somewhat

† Millfeed is a general term applied to any one of the fractions of the kernel left after flour removal. Shorts is the fraction of millfeed left after removal of coarse bran and most of the red dog and germ. Millrun is the total residue left after flour removal.

Table 13.8. *Yield and analysis of flour produced by milling and sifting of wheat shorts*†

% yield	% protein (N × 5.7)	% fat	% fiber	% ash
40.5	23.4	6.8	2.0	3.5

† Shorts at 9 % moisture before milling. Flour analysis on dry basis.

depending on the moisture and protein content of the original millfeed, but Table 13.8 indicates a typical analysis of such a flour.

Lysine content of the protein concentrate is normally 4–5 g/16 g N. The PER value, about 1.9 compared to 2.5 for casein, is much superior to that of regular wheat flour (Fellers *et al.*, 1968). The nutritive value of the protein is also superior to that of regular flour, and almost as good as wheat germ protein (Miladi *et al.*, 1972).

Compared to the original millfeeds, the protein concentrates contain more starch, protein, fat, thiamine, folic acid and choline, but less riboflavin, vitamin B_6, niacin, pantothenic acid, and fiber. However, all vitamins present were higher than in white flour. Generally the protein concentrates prepared from millfeeds may be described as natural wheat food products, high in B vitamins and good-quality protein, and of moderate fiber content. Commercially these concentrates, or similar ones produced by various modifications, have achieved considerable success in exports to less developed countries (Pence, 1968).

Protein concentrates can also be prepared from wheat millfeeds by a wet process similar to that described in the sections on rice and oats. The millfeed is extracted with alkali at pH 9 then squeezed to separate the liquid phase. From this juice, by ethanol precipitation, heat coagulation or isoelectric precipitation, fractions rich in protein, starch and fat, and low in fiber are obtainable (Fellers *et al.*, 1966*b*; Saunders *et al.*, 1972). Table 13.9 shows typical concentrates prepared by this process (Saunders *et al.*, 1972). These protein concentrates contain good quality protein, high lysine levels of about 5 g/16 g N, and polyunsaturated fats. Further investigations are necessary, however, to assess their commercial potential.

Triticale and rye

The wet alkaline process to manufacture protein concentrates from wheat millfeeds can also be successfully applied to milling products of triticale

Table 13.9. *Yield and analysis of concentrates prepared from wheat millfeeds by a wet process†*

Millfeed	% yield	% protein (N × 6.25)	% fat	% fiber	% ash	% starch
Shorts	20.5	43.1	11.7	0.8	4.4	39.4
Bran	13.0	34.7	9.5	0.8	4.3	40.6
Millrun[a]	18.5	26.2	7.0	0.8	4.1	47.7
Shorts[b]	9.9	65.3	17.6	0.5	3.5	< 1

† Dry basis. Products prepared by heat coagulation. [a] The whole wheat kernel *less* 75 %-extraction white flour. [b] Starch removed before isoelectric precipitation of concentrate.

Table 13.10. *Yield and analysis of concentrates prepared from rye and triticale milling products by a wet process†*

Millfeed	% yield	% protein (N × 6.25)	% fat	% fiber	% ash	% starch
Triticale coarse bran	13.9	47.7	11.8	0.3	5.3	16.7
Triticale fine bran	15.0	45.1	10.1	0.3	5.5	17.7
Rye middlings	22.5	33.2	3.9	0.5	4.8	42.1

† Dry basis. Products prepared by heat coagulation.

(*Triticale*) and rye (*Secale cereale*) (Saunders *et al.*, 1972). Table 13.10 lists yields and analysis of concentrates obtainable by this process. Further investigations are necessary to assess commercial potential.

Sorghum

Starch is prepared by wet milling sorghum (*Sorghum vulgare subglabrescens*) (Wall & Ross, 1970) but the yields of protein-rich byproducts have not been published. The figures in Table 13.11 (Watson, 1970) are from laboratory millings. On a laboratory scale, fine grinding and air classification of sorghum fractions has been shown to concentrate protein in the fine particles. From different sorghums, grits produced concentrates having protein contents up to 21.6 %, flour endosperm gave concentrates containing up to 19.3 % protein, and horny endosperm gave concentrates containing up to 22.0 % protein (Stringfellow & Peplinski, 1966). After removal of the bran and germ, it is possible to abrade from the kernel endosperm a fraction containing 18 % protein, representing a quarter of the original kernel weight (Normand *et al.*, 1965).

Table 13.11. *Yield and protein content of protein concentrates obtained in wet milling of sorghum*†

	% yield	% protein (N × 5.7)
Whole grain	100	14.1
Solubles	6.9	48.2
Gluten	10.6	46.7

† Dry basis.

Barley

Abrasive milling of pearled barley (*Hordeum vulgare*) kernels (10–12 % protein) can provide a flour fraction in 9–10 % yield containing 18–24 % protein. Air classification of barley flours into high protein fractions has been studied and the potential use of the various fractions evaluated. The actual protein enrichment was relatively small; the most attractive concentrate obtained was enriched only by a factor of about 14 % over the original flour protein level, and the yield was less than 10 % (Pomeranz *et al.*, 1971).

It is obvious that cereals can be utilized to produce concentrates enriched in protein. Air classification of flours appears to be most attractive commercially, although the yields of enriched fractions are generally low. Protein concentrates by wet processing of cereal bran residues are the most attractive from the viewpoint of yield, protein content, overall composition and protein nutritive value.

References

Chen, L. & Houston, D. F. (1970). Solubilization and recovery of protein from defatted rice bran. *Cereal Chemistry*, **47**, 72.

Cluskey, J. E., Wu, Y. V., Wall, J. S. & Inglett, G. E. (1973). Oat protein concentrates from a wet milling process: preparation. *Cereal Chemistry*, **50**, 475.

Fellers, D. A., Shephard, A. D., Bellard, N. J. & Mossman, A. P. (1966*a*). Protein concentrates by dry milling of wheat millfeeds. *Cereal Chemistry*, **43**, 715.

Fellers, D. A., Shephard, A. D., Bellard, N. J., Mossman, A. P. Johnston, P. H. & Wasserman, T. (1968). Protein concentrates by dry milling of wheat millfeeds. II. Compositional aspects. *Cereal Chemistry*, **45**, 520.

Fellers, D. A., Sinkey, V., Shephard, A. D. & Pence, J. W. (1966*b*). Solubilization and recovery of protein from wheat millfeeds. *Cereal Chemistry*, **43**, 1.

Houston, D. F. & Kohler, G. O. (1970). *Nutritional properties of rice*. US National Academy of Sciences, Washington, DC.

Houston, D. F. & Mohammad, A. (1966). Air-classification and sieving of rice bran and polish. *Rice Journal*, **69**, 20.

Houston, D. F., Mohammad, A., Wasserman, T. & Kester, E. B. (1964). High-protein rice flours. *Cereal Chemistry*, **41**, 514.

Inglett, G. E. (1970). *Corn: Culture, processing, products*. AVI Publishing Co., Westport, Connecticut.

Johnston, P. H. & Fellers, D. A. (1971). Process for protein–starch separation in wheat flour. 2. Experiments with a continuous decanter-type centrifuge. *Journal of Food Science*, **36**, 649.

Kent, N. L. & Evers, A. D. (1969). Fine-grinding and air-classification of subaleurone endosperm of high protein content. *Cereal Science Today*, **14**, 142.

Lew, E. J., Houston, D. F. & Fellers, D. W. (1970). Extraction of protein from full-fat rice bran. *Cereal Science Today*, **16**, Abstract 92.

Miladi, S., Hegsted, D. M., Saunders, R. M. & Kohler, G. O. (1972). The relative nutritive value, amino acid content, and digestibility of the proteins of wheat mill fractions. *Cereal Chemistry*, **47**, 119.

Normand, F. L., Hogan, J. T. & Deobald, H. J. (1965). Protein content of successive peripheral layers milled from wheat, barley, grain sorghum, and glutinous rice by tangential abrasion. *Cereal Chemistry*, **42**, 359.

Pence, J. W. (1968). New wheat foods. *Nutrition Review*, **26**, 291.

Pomeranz, Y., Ke, H. & Ward, A. B. (1971). Composition and utilization of milled barley products. I. Gross composition of roller-milled and air-separated fraction. *Cereal Chemistry*, **48**, 47.

Reiners, R. A., Pressick, J. C., Urquidi, R. L. & Warnecke, M. O. (1972). Corn protein concentrates for food use. In: IXth International Congress on Nutrition, Mexico.

Saunders, R. M., Conner, M. A., Edwards, R. M. & Kohler, G. O. (1972). Wet processing of wheat millfeeds. *Cereal Science Today*, **17**, Abstract 66.

Shroder, J. D. & Heiman, V. (1970). Feed products from corn processing. In: *Corn: Culture, processing products* (ed. G. E. Inglett), p. 220. AVI Publishing Co., Westport, Connecticut.

Stringfellow, A. C. & Peplinski, A. J. (1966). Air classification of sorghum flours from varieties representing different hardnesses. *Cereal Science Today*, **11**, 438.

Wall, J. S. & Ross, W. M. (1970). *Sorghum production and utilization*. AVI Publishing Co., Westport, Connecticut.

Watson, S. A. (1970). Wet milling process and products. In: *Sorghum production and utilization* (ed. J. S. Wall & M. W. Ross). AVI Publishing Co., Westport, Connecticut.

Wu, Y. V., Cluskey, J. E., Wall, J. S. & Inglett, G. E. (1973*a*). Oat protein concentrates from a wet milling process: composition and properties. *Cereal Chemistry*, **50**, 481.

Wu, Y. V. & Stringfellow, A. C. (1973*b*). Protein concentrates from oat flours by air classification of normal and high protein varieties. *Cereal Chemistry*, **50**, 489.

Ziegler, E. & Greer, E. N. (1972). Principles of milling. In: *Wheat: chemistry and technology* (ed. Y. Pomeranz), p. 115. American Association of Cereal Chemists, St Paul, Minnesota.

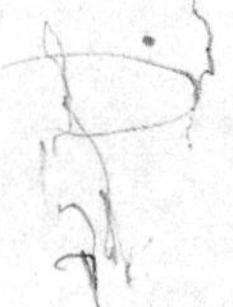

14. Leaf protein

N. W. PIRIE

General principles

Work on leaf protein (LP) proceeded more fully in accordance with the basic principles of IBP than work on the other protein sources discussed in this synthesis volume. It formed part of IBP programs in India, New Zealand, Nigeria, Sweden, Uganda and the United Kingdom, and was the subject of a Working Group meeting from which Handbook 20 resulted. Relevant work not specifically mentioned in IBP programs is being, or has been, done in Australia, Brazil, Canada, Eire, France, Hungary, Jamaica, New Guinea, Pakistan, Sri Lanka and the USA.

The history of LP production is outlined in IBP Handbook 20 and need not be repeated here. The main reasons for thinking that it would sometimes be advantageous to fractionate a leafy crop instead of using the crop as fodder, or using the land to grow a conventional crop yielding an edible seed or underground part, are:

(1) Between 10 and 30 % of the protein in a forage is converted into human food by ruminants, whereas 40 to 60 % of the protein can be extracted. The approximate consequences of fractionating a forage crop rather than using it as fodder are shown in Fig. 14.1.

(2) Leaves are the main site of protein synthesis and there are losses during translocation to other parts of a plant.

(3) When LP is made, the crop is harvested when less mature than when silage is made, and much less mature than when hay is made or a conventional crop is taken; the cost of harvesting is greater but an immature crop is not at risk for so long from diseases and pests.

(4) Crops that regrow several times after being cut young, or perennial crops, maintain cover on the ground; this enables fuller use to be made of sunlight and protects the ground from erosion.

(5) The fibrous residue contains the protein that was not extracted. Depending on the processing conditions, it can have two to five times as great a percentage of dry matter as the original crop and can therefore be dried to produce conserved ruminant feed economically.

Points (1) and (2) were the first to attract the attention of scientists and led to small-scale research being started in many countries. Industrialists,

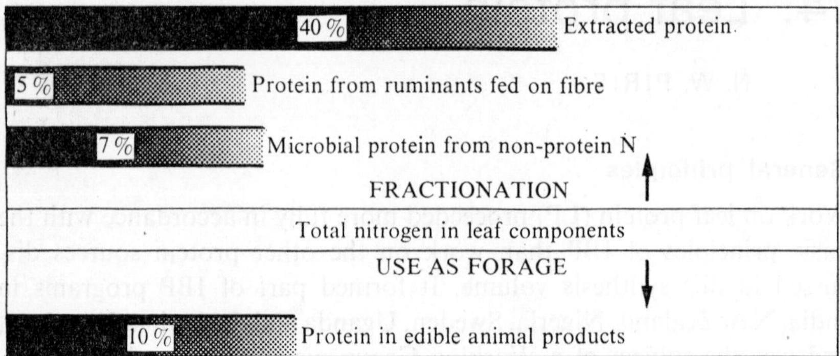

Fig. 14.1. Differences in protein yield between a crop used as forage or fractionated to extract the protein.

and those responsible for national and international research policy, remained sceptical; they were uncertain about the acceptability of LP as a human food, and they assumed, in spite of the experience of all those who worked on the process, that extraction would be very expensive. Although point (5) has repeatedly been stressed (e.g., Pirie, 1942, 1951, 1966) its importance was only recently recognised. Large-scale work done in various parts of the world is primarily directed towards making ruminant fodder that can be economically conserved – the extracted protein is regarded essentially as a by-product to be used for feeding pigs and poultry. Those concerned with nutrition in countries where malnutrition is common, especially those in the wet tropics where points (3) and (4) are important, tend to look on this diversion of interest from human to animal feeding as unfortunate. The diversion has, however, the merit that it will supply figures for the processing costs and so eliminate one argument against the mechanical processing of leaves. That argument has a long history. Lawes (1885) wrote 'It might be possible by some chemical process to produce from grass a nutritious substance which a man could use as food, but the food so extracted would be far more costly than as it existed in the grass, and no one would think of preparing such a food for oxen or sheep'.

Crops

Species and varieties selected for seed production or for a use other than LP extraction have been the source of most of the LP made in bulk. If varieties, possibly of species not at present used in agriculture, were investigated, yields would probably be greater than those so far attained. Nevertheless, annual yields are already impressive. Arkcoll & Festenstein (1971), in the UK without irrigation, get 2 t/ha from a succession of crops, including winter wheat (*Triticum aestivum*), fodder radish (*Raphanus sativus*) and mustard (*Sinapis alba*). Irrigated lucerne (*Medicago sativa*) in New Zealand (Allison & Vartha, 1973) gives 1.95 t/ha. In Aurangabad (Dev *et al.*, 1974; Deshmukh *et al.*, 1974) many different species were compared at different times of year and with different manurial treatments. Cowpea (*Vigna unguiculata*) gave 895 kg/ha in 80 days i.e., more than 4 t if that rate could have been maintained for a year. In short term experiments yields as great as 17 kg of extracted protein per ha per day have been claimed.

Several principles have to be borne in mind in selecting species for study. Protein extracts more readily from soft lush leaves than from those that are fibrous and dry; even when pulped with added alkali, acid leaves do not extract so well as those that give neutral extracts; leaves that give glutinous or slimy extracts are difficult to handle. It is obviously necessary to use leaves that can be harvested mechanically and advantageous to use a species that will regrow after cutting; this probably excludes tree leaves, though coppiced trees have potentialities (Pirie, 1968). Equally obviously, mixed weeds from untended ground are useless – if the growth could be harvested mechanically and is being manured to ensure an adequate yield, it would be better to use the ground to grow a desirable species. Water weeds are an exception to that generalisation (Boyd, 1971); they often grow luxuriantly, but little is known about the extractability of the protein in them. Good yields of protein have come from leaves that are by-products, e.g., beans (*Phaseolus* spp. and *Vicia faba*), jute (*Corchorus* spp.), peas (*Pisum sativum*), ramie (*Boehmeria nivea*), potato (*Solanum tuberosum*) and sugar beet (*Beta vulgaris*).

Separation of extract from fibre

The yields given above were measured on 4 to 5 kg samples of leaf taken from within a crop, pulping them in the unit designed for IBP (Davys & Pirie, 1969), pressing a sample in the unit similarly designed (Davys, Pirie & Street, 1969) and measuring the amount of protein precipitable from the extract with trichloroacetic acid. In large-scale work it is usually advantageous to re-extract the fibre; this can give half as much protein again as a single extraction, but it would be difficult in the laboratory to get quantitative and repeatable results from a double

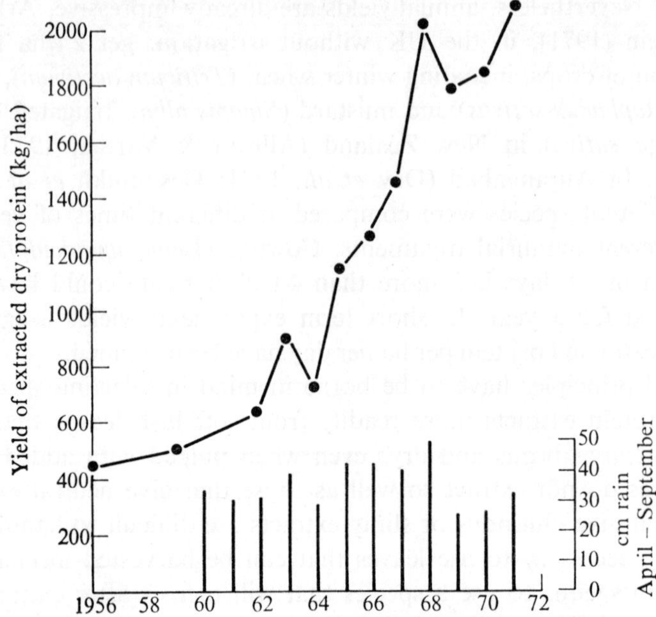

Fig. 14.2. Improvements in the annual yield of extracted leaf protein at Rothamsted Experimental Station.

extraction. The manner in which increasing skill in agronomy and processing have increased yields at Rothamsted are shown in Fig. 14.2.

Opinions differ about the best design of extraction unit for large-scale work. At Rothamsted we have been improving the design of the fixed hammer, unchokable pulper described in IBP Handbook 20 and remain convinced that its basic design is preferable to any of the alternatives that have been tried. We are more open minded about the best type of press. Our belt press (Davys & Pirie, 1965) does not express juice as fully as is possible and it may be advantageous to press again in a more robust press if the fibre is finally to be dried (Pirie, 1973). Because the concentration of protein in the juice that comes out with gentle pressure is three or four times as great as in the juice coming out on intense pressing, such pressing is not important for protein production.

For commercial production in USA, sugar cane rolls are used (see p. 142). The 3-roll mill is an admirable device for crushing a hollow tube – the job for which it was designed. It is poorly adapted for rubbing leaves in the manner that releases juice from them, and, unless the rolls revolve extremely slowly, it does not maintain pressure for long enough to allow the juice to flow away from the fibre. To get satisfactory liberation of juice, many passages through a 3-roll mill are therefore needed. This not only wastes power but necessitates prolonged contact between the leaf pulp and the liberated juice; that diminishes protein yield (Davys, Pirie & Street, 1969), allows more opportunity for deleterious complex formation between protein and polyphenols and similar substances in leaf extracts, and leads to loss of β-carotene. It is claimed that crops can be pulped satisfactorily by forcing the mass through 1 cm holes in a die.

The cost of heavy-duty reduction gears gives high-speed (800 to 1700 rev/min) pulpers such as those used at Rothamsted a seeming advantage, but half the power they consume is wasted in wind and friction. Now that slow-speed hydraulic drives are becoming widely used, it would be worth while trying to develop a slow unit (Pirie, 1971) able to handle several tonnes an hour but holding only a few kilograms of crop at any instant. These mechanical points deserve emphasis because, after the cost of the crop itself, pulping is the most expensive part of the process. During the past 12 years our power consumption per kilogram of extracted protein has been diminished to $\frac{1}{4}$; the limit to improvement has not been reached.

Separation of protein from extract

Heat coagulation is generally accepted as the most satisfactory method for making a protein curd. Green, predominantly 'chloroplastic', protein coagulates at 50 to 60 °C; if that is separated, colourless 'cytoplasmic' protein separates at 70 °C. No more protein coagulates on further heating, but heating to 100 °C is probably advantageous in other ways; it ensures a more nearly sterile product and it inactivates leaf enzymes more completely. When steam is injected into a stream of juice, heating takes place in 1 or 2 seconds; this produces a hard, easily filtered curd, and there is less enzyme action before inactivation. Chlorophyllase-rich plants such as lucerne and wheat show the importance of this; the chlorophyll in LPs made by heating to 80 °C was almost completely hydrolysed to chlorophyllide, whereas there was little hydrolysis during quick heating to 100 °C (Arkcoll & Holden, 1973). When a slowly heated protein is washed in slightly acid conditions, pheophorbide is formed by loss of magnesium. LP made by heating lucerne juice slowly (Hove *et al.*, 1973) contained enough pheophorbide to photosensitise rats fed on it (Lohrey *et al.*, 1973).

After heat coagulation the protein can be separated from the 'whey' and dried or preserved by the standard methods of industrial chemistry. The 'whey' must be disposed of to prevent local pollution. The simplest course is to put it back on the land where the NPK in it would be useful and the 1 to 3 % of carbohydrate is a soil conditioner. Ideally it would be used as a medium for culturing micro-organisms (Pirie, 1971).

Feeding value

Woodham & Singh surveyed in IBP Handbook 20 (Pirie, 1971) the results of animal and human feeding experiments. Properly made LP is better than any seed protein but not as good as milk or egg – this would be expected from its amino acid analysis. Although most preparations contain more than 2 g methionine for 16 g N, rat growth has always been increased by adding more methionine to the diet. Part of the methionine in LP is probably unavailable. There is no evidence that this is the consequence of complex formation or oxidation during the isolation, but the possibility should be borne in mind and preparations, made carefully and quickly, should be tested to establish the intrinsic merits of LP. Such species differences as have been observed between preparations can plausibly be explained as the result of different amounts of damage done to them during preparation.

References

Allison, R. M. & Vartha, E. W. (1973). Yields of protein extracted from irrigated lucerne. *New Zealand Journal of Experimental Agriculture*, **1**, 35.

Arkcoll, D. B. & Festenstein, G.N. (1971). A preliminary study of the agronomic factors affecting the yield of extractable leaf protein. *Journal of the Science of Food and Agriculture*, **22**, 49.

Arkcoll, D. B. & Holden, M. (1973). Changes in chloroplast pigments during the preparation of leaf protein. *Journal of the Science of Food and Agriculture*, **24**, 1217.

Boyd, C. E. (1971). Leaf protein from aquatic plants. In: *Leaf protein: its agronomy, preparation, quality and use*. IBP Handbook 20 (ed. N. W. Pirie), p. 44. Blackwell Scientific, Oxford.

Davys, M. N. G. & Pirie, N. W. (1965). A belt press for separating juices from fibrous pulps. *Journal of Agricultural Engineering Research*, **10**, 142.

Davys, M. N. G. & Pirie, N. W. (1969). A laboratory-scale pulper for leafy plant material. *Biotechnology and Bioengineering*, **11**, 517.

Davys, M. N. G., Pirie, N. W. & Street, G. (1969). A laboratory-scale press for extracting juice from leaf pulp. *Biotechnology and Bioengineering*, **11**, 528.

Deshmukh, M. G., Gore, S. B., Munikar, A. M. & Joshi, R. N. (1974). The yields of leaf protein from various short-duration crops. *Journal of the Science of Food and Agriculture*, **25**, 717.

Dev, D. V., Batra, U. R. & Joshi, R. N. (1974). The yields of extracted leaf protein from lucerne (*Medicago sativa* L.). *Journal of the Science of Food and Agriculture*, **25**, 725.

Hove, E. L., Lohrey, E., Urs, M. K. & Allison, R. M. (1974). The effect of lucerne-protein concentrate in the diet on growth, reproduction and body composition of rats. *British Journal of Nutrition*, **31**, 147.

Lawes, J. B. (1885). Sugar as a food for stock. *Journal of the Royal Agricultural Society* (2nd Ser.) **21**, 81.

Lohrey, E., Tapper, B. & Hove, E. J. (1974). Photosensitization of albino rats fed lucerne protein concentrate. *British Journal of Nutrition*, **31**, 159.

Pirie, N. W. (1942). Direct use of leaf protein in human nutrition. *Chemistry and Industry*, **61**, 45.

Pirie, N. W. (1951). The circumvention of waste. In: *Four thousand million mouths* (ed. F. LeGros Clark & N. W. Pirie), p. 180. Oxford University Press, London.

Pirie, N. W. (1966). Fodder fractionation: an aspect of conservation. *Fertiliser and Feeding Stuffs Journal*, **63**, 119.

Pirie, N. W. (1968). Food from the forests. *New Scientist*, **40**, 420.

Pirie, N. W. (1971). *Leaf protein: its agronomy, preparation, quality and use.* IBP Handbook 20. Blackwell Scientific, Oxford.

Pirie, N. W. (1973). Plants as sources of unconventional protein foods. In: symposium on *the Biological Efficiency of Protein Production* (ed. J. G. W. Jones), p. 101. Cambridge University Press, London.

15. Industrial production of leaf protein in the USA

G. O. KOHLER & E. M. BICKOFF

While the concepts involved in the development of leaf protein products suitable for direct consumption by human beings, or for use as a feed protein source are most attractive, their practical application proved exceedingly elusive. The US Department of Agriculture industrial team was set up to pursue the possibility of production of animal grade leaf protein concentrate (LPC) (Phase I of the program) as a by-product of dehydrated alfalfa production based on the following ideas: (1) The color and flavor problem of whole green LPC is not of importance in an animal feed. (2) A market for xanthophylls as a source of pigment in broiler and laying hen rations had developed. (3) Computer systems for the determination of ascribed dollar value of a new potential feed ingredient lessened the risk of introducing new products to the market. (4) The logistics of growing, harvesting, and handling large volumes of fresh young alfalfa had been developed by the forage dehydration industry. Dehydration plants operating on a 24-hour per day basis throughout the growing season reduced fixed costs of operating to a minimum. (5) Adding a dejuicing step to the existing dehydration plant could reduce overall energy usage by 25–30 % (Kohler *et al.*, 1973). (6) By operating a light pressing operation (e.g., with sugar cane rolls), the leaf protein could be produced as a by-product of high quality dehydrated alfalfa or grass, permitting market development on the protein to be carried out at minimum risk. Increased yields could be obtained by the use of prepulping and/or a twin-screw press if economic analysis showed it to be desirable. (7) After establishing an economically viable plant for producing by-product LPC as a high value feed for poultry (the PRO-XAN I process), the next phase of the research would be to insert unit operations in the process to split the protein of the juice into the chloroplastic and the soluble (white) protein fractions (Phase II or PRO-XAN II process).

During the past several years, the increasing cost of protein sources (soybean meal and casein) has further improved the economic potential of LPC.

The Phase I research is essentially complete. A large commercial dehydration company has been producing by-product LPC commercially for about four years.† By omitting a grinding step, a pressed alfalfa is obtained which can be more effectively dehydrated in a conventional dryer, as compared with shredded alfalfa containing small easily-burned fibers. The choice of sugar cane rolls was made for the commercial venture since the 35 % juice yield obtained with these rolls leaves a dehydrated alfalfa product which can be sold with standard guarantees of protein and carotene contents. Such guarantees can be maintained by the use of immature high grade alfalfa as a starting material. The 17 % protein alfalfa produced in this plant is fully equivalent in nutrients to 17 % protein alfalfa produced from lower grade starting material (Kohler *et al.*, 1973). In the present plant operation, the residual brown juice, after protein removal, is added to the pressed alfalfa. Hence, the energy conservation potential of the process is not fully realized commercially.

Kohler *et al.* (1968, 1973) give key references to publications containing the details of the PRO-XAN I process. The alkaline conditions (pH 8 to 8.5) used diminish protease and lipoxygenase activity. It is known that such alkaline conditions also prevent loss of magnesium from chlorophyll. It will be interesting to determine whether chlorophyllase activity is reduced under these conditions.

Recently, the conditions for the efficient separation of the green and white fractions of leaf protein have been established on a laboratory scale (de Fremery *et al.*, 1973). In addition, the scale-up to a pilot plant has now been reported (Edwards *et al.*, 1973).

Present research is directed to purification of the white protein fraction of leaf protein in undenatured form. There is a great deal of industrial interest in the US in both feed and food LPC, and it seems most likely that the next several years will see further commercial development.

Reference to a company and/or product named by the Department is only for purposes of information and does not imply approval or recommendation of the product to the exclusion of others which may also be suitable.

References

de Fremery, D., Miller, R. E., Edwards, R. H., Knuckles, B. E., Bickoff, E. M. & Kohler, G. O. (1973). Centrifugal separation of white and green protein fractions from alfalfa juice following controlled heating. *Journal of Agricultural and Food Chemistry*, **21**, 886.

† Batley–Janss Enterprises, Inc., Brawley, California, sold this product under the trade name 'X-Pro®'.

Edwards, R. H., Miller, R. E., de Fremery, D., Knuckles, B. E., Bickoff, E. M. & Kohler, G. O. (1973). The production of edible white protein from alfalfa. Presented at the 166th ACS meeting, Chicago, Illinois, August 26–31. Abstracts of Papers, p. AGFD, Paper No. 78.

Fomin, V. I. (1973). Mechanical dewatering of lucerne by repeated pressing. Proceedings of the First International Crop Drying Congress, Oxford, p. 341.

Kohler, G. O., Bickoff, E. M. & de Fremery, D. (1973). Mechanical dewatering of forage and protein byproduct recovery. Proceedings of the First International Green Crop Drying Congress, Oxford, p. 326.

Kohler, G. O., Bickoff, E. M., Spencer, R. R., Witt, S. C. & Knuckles, B. E. (1968). Wet processing of alfalfa for animal feed products. Proceedings of the Tenth Technical Alfalfa Conference (held at Reno, Nevada, July 11, 1968), ARS-74-46, p. 71.

Knuckles, B. E., Bickoff, E. M. & Kohler, G. O. (1972). PRO-XAN process: methods for increasing protein recovery from alfalfa. *Agricultural and Food Chemistry*, **20**, 1055.

PART III

Concentrates made by biological conversion

16. Protein from non-domesticated herbivores

K. L. BLAXTER

The amount of animal protein which populations obtain from non-domesticated species is not precisely known, though FAO in its annual tabulations of food consumption in different countries estimates the amount of 'other meat' consumed in each country. It is obvious that in many rural communities which hunt wild animals, such meat is a major source of protein, but even in urban areas meat from wild animals is marketed as a relished food. This is so in Ghana, a country not renowned for its wild life (Asibey, 1966), and equally in the more developed countries of Europe and North America. Thus, some 5000 tonnes of meat from wild red deer is imported by continental Europe from New Zealand and Scotland each year to augment that obtained from internal sources. In the developed countries meat from game birds and animals is, however, more of a luxury item of diet than a major or essential source of protein.

The possibility that some large wild herbivorous mammals might have attributes which make them eminently suitable in some specific contexts as major sources of meat, has occurred to many, and IBP has promoted a number of studies to supplement a considerable programme of investigations sponsored or supported by other agencies (Golley & Buechner, 1968). The arguments adduced in favour of meat production from wild species are mainly ecological and are concerned with the relative productivity of wild species and imported domestic ones in particular habitats, most of which are natural grasslands or scrublands hitherto not extensively subjected to agrarian or pastoral development. The ecological arguments have been made by, among others, Talbot (1963), Harthoorn (1968), Crawford (1968*c*), de Vos (1969), Hopcraft (1970), Kyle (1972), and perhaps most convincingly by Darling (1960) who concluded: 'Only under the natural communities of game animals can a high biological capture and turnover of solar energy be maintained.' Certainly, as Kay (1970) has pointed out, in some habitats, notably very arid areas and barren northern wastes, there is no question of using domesticated species because unlike wild species they are not adapted to the climatic and nutritional conditions such habitats impose.

Skinner's (1970) view on the more general issue is that 'without in any way wishing to detract from the contribution made by zoologists it is a fact that their enthusiasm for and inadequate scientific evidence has too often led to the case for meat production being overstated'. It is indeed true that at present few direct long-term comparisons of the productivity of wild animals and that of domesticated livestock in the same habitat have been made and the question of their relative merits is still open. Nevertheless, much knowledge has accrued about the physiology, population dynamics, nutritional demands and body composition of some wild species not only to enable their potential to be more adequately assessed but also to facilitate indirect comparison of their productivity with that of domesticated stock.

Meat from wild and domesticated ruminants

Dressed carcase is the skinned, gutted carcase with the head removed at the occiput and those parts of the leg distal to the radio-ulna and the tibio-fibula discarded, and the proportion is usually referred to as the killing-out percentage. Table 16.1 shows mean weights of carcases of some of the ruminant species which have been seriously considered as economic producers of meat. This table confirms with a wider range of species the conclusions drawn from studies in East Africa (Ledger, 1968) that the carcase yield of many wild ruminants is slightly higher than that of domesticated species. Exceptions are the African buffalo and the American bison. It appears to be a characteristic feature of many of the wild ruminants, particularly sub-tropical ones, that their killing-out percentages are relatively stable, unaffected greatly by age and sex.

The yield of deboned meat from the dressed carcases of wild animals at 80 to 85 % is similar to that from domesticated species, but there is a tendency for the amount of meat in the hind quarter of the carcase to be greater in the wild species than in the domesticated ones. Table 16.2 summarises on a comparative basis data for the red deer in which it is apparent that the proportional yield of first quality meat from the back, haunch and shoulder is greater than in sheep and cattle dissected using similar techniques.

The yield of deboned meat from the carcase and its distribution is, however, less striking than the relative proportions of lean meat and dissectible fat. Ledger's series of dissections (Ledger, 1963, 1968) show this well. The carcases of zebu (*Bos indicus*) steers in East Africa contained 28.6 ± 4.6 % of fat, while eland (*Taurotragus oryx*) contained

Table 16.1. *The liveweights of some domesticated and wild ruminants and their dressed carcases as percentages of liveweight*†

	Mean liveweight (kg)	Dressed carcase (% liveweight)	Reference
Domesticated species			
Cattle‡	500	55	
Sheep‡	40	47	
Goat	30	47	Devendra & Burns (1970)
Horses (Russian Kazakhstan meat horses)	380	50	Venjarskij (1959, 1963)
Wild species – cervids			
Red deer (European) (*Cervus elaphus*)	60	59	Fowler (1973), Mitchell (1972)
Antilocaprids			
North American Pronghorn antelope (male) (*Antilocapra americana*)	54	56	Mitchell (1971)
Bovids			
Common eland (*Taurotragus oryx*)	508	59.1	Ledger (1968)
African (Cape) buffalo (*Syncerus caffer*)	589	49.6	Young & van den Heever (1969)
American bison (bulls 6 yr) (*Bison bison canadensis*)	577	54.1	Halloran (1957)
Wildebeest (*Connochaetes taurinus*)	230	57.7	Young *et al.* (1969)
Springbok (*Antidorcas marsupialis*)	36	57.9	van Zyl *et al.* (1969), Skinner *et al.* (1971)
Impala (*Aepyceros melampus*)	45	57.4	Young & Wagener (1968)

† An extended tabulation of data obtained with 32 African species of ungulate is given by von la Chevallerie (1970).

‡ Average values: the % increases with fatness.

only 4.2 ± 2.4 %, and impala 2.0 ± 1.3 %. The same is true of temperate wild ruminants; the meat is characteristically lean, though fat content varies more and seasonally. Table 16.2 includes a comparison of the amounts of fat in the deboned carcase of red deer, sheep and cattle which indicates large differences between species and the last column shows that even the muscles themselves, dissected free of intermuscular fat, contain less fat in deer than in cattle and sheep. It is particularly noteworthy that on a liveweight basis the yield of first quality meat with a low fat content is considerably greater in red deer than it is in sheep kept on similar land. Generally it seems that the yield of lean tissue per unit liveweight is likely to be greater in wild species than in domesticated cattle, zebu and sheep.

The composition of the lean meat of wild species has been studied systematically by Mann (1964) and Crawford (1968*a*). The amino acid

Table 16.2. *The yield and gross composition of the carcases of well-fed red deer, cattle and sheep*†

Species	Body weight (kg)	Dressed carcase weight (DCW) (% body weight)	Dissected subcutaneous fat of carcase (% DCW)	Dissected abdominal and peri-renal fat of carcase (% DCW)	Bone-free first quality meat (% DCW)	Bone-free second quality meat (% DCW)	First quality meat (% of liveweight)	Extractable (chemical) fat in carcase less that in dissected fat (%)	Extracted fat in muscle *per se* (intra-muscular fat) (%)
Deer (6 months)	48	59.1	3.6	2.6	56.1	22.2	33.1	6.5	1.9
Deer (3 years)	58	53.4	4.3	1.8	57.4	22.6	30.7	8.7	2.4
European									
Cattle	500	55.0	8.2	3.5	45.6	28.5	25.1	22.1	5.5
Sheep	41	47.1	18.2	3.7	38.7	23.2	18.2	22.3	6.8

† Based on carcase dissections and compilations made by Dr V. Fowler, Mr G. A. M. Sharman & Dr R. N. B. Kay during studies of the red deer for an IBP project (V. Fowler, 1973, unpublished data).

compositions of the mixed proteins of muscular tissue from a variety of wild and domesticated species do not differ significantly from one another, and any small differences in contents of B-complex vitamins noted by Mann may well be due, as he states, to storage and sampling difficulties. Few studies of the traits associated with the quality of the meat from wild species have been made. Colour, tenderness and flavour tests on seven wild South African ungulates by von la Chevallerie (1972) show, however, real differences, with springbok meat being preferred. These results thus suggest that lean meat from wild ruminants probably resembles lean meat from domesticated species in terms of its nutritive value and, within limits, its acceptability. The meat obtained from wild species has been shown to contain larger proportions of polyunsaturated acids in the total fatty acids than meat from domesticated ones or of wild species kept in parks (Crawford, 1968*b*). This may largely reflect the increased deposition of saturated and mono-unsaturated acids as triacylglycerols in domesticated and confined animals (Garton, 1969; Crawford *et al.*, 1970); indeed Crawford *et al.* comment on the inverse relation between the percentage lipid in meats and the percentage of polyunsaturated acids in this lipid.

A further aspect of meat from wild animals relates to its safety for man; that is its freedom or otherwise from organisms pathogenic to him. Studies by Young *et al.* (1969) and Young & van den Heever (1969) on African buffalo and wildebeest indicate the presence of a number of parasites in edible tissues including *Cysticercus regis* (which is not transmissible to man) which could result in the condemnation of parts of carcases on grounds of aesthetic acceptability; the meat is 'measly'. They also point out that these wild animals can well harbour diseases common to man, notably anthrax, wesselsbron disease, brucellosis and taeniasis. In addition, spoilage organisms, and in particular *Salmonellae*, could infect carcases, particularly if they are poorly handled. In this respect the methods of harvesting wild animals as food are pertinent. The shooting of a wild animal can be done in such manner that the carcase is not affected. It is estimated, however, that 10 % of the meat from wild African species is spoiled even when animals are shot from hides by experienced hunters using high velocity rifles. This meat has to be discarded (von la Chevallerie, 1970). In Scotland some 40 % of the prime cuts of red deer shot in the wild (corresponding to first-class meat in Table 16.2) is probably spoiled by inexpert shooting (Bland, 1969). To this loss must be added the loss occasioned by animals that are wounded and escape. This has been variously estimated to be about 10 %.

Such questions of disease control to safeguard public health, and the problems of loss of meat as a result of harvesting methods, are relevant to the two alternative policies to be adopted in the use of wild species of herbivorous mammal as sources of meat. One consists of the planned cropping and management of a natural population, the other a partial or complete domestication of the species, so that surplus animals can be killed at some central place and their carcases handled hygienically.

The extent to which a wild population can be cropped for meat reveals some curious features. The results of game eradication programmes for tsetse fly control, suggest that the numbers of some species – notably wart hog (*Phacochoerus aethiopicus*) and duiker (*Sylvicapra grimmia*) increased and that it was only zebra, giraffe and rhinoceros which were eliminated (Riney, 1964). In some National Parks a sustained cropping to prevent overgrazing by wild species has to take place, yet many of the natural populations of ungulates maintain stable numbers without their habitat becoming grossly eroded. Harthoorn (1968) has argued that predation may be responsible; this term includes poaching. The extent to which a population can be cropped is certainly less than the optimistic theoretical estimates made by Talbot *et al.* (1965) from reproduction rates, namely 25 % of the total population every year for a monotocous species breeding annually. The incidence of infertility and pre- and post-natal mortality in many species are factors which signally reduce theoretical estimates. Thus, studies of North American deer (*Odocoileus* and *Cervus* spp.) using age-specific birth and mortality rates, indicate that a 20 % harvest every second year would maintain the population stable in number (Walters & Bandy, 1972). For many species similar statistics are not available but it may well prove that an annual cropping rate of 10 to 15 % is a more reasonable expectation. A low cropping rate and a low animal density (animals/km^2), such as would occur in marginal habitats, clearly creates considerable technical problems in harvesting wild meat on a commercial basis.

An alternative is to domesticate certain selected species so that they can be handled more readily and subjected to 'improvement' in the sense that agricultural practices can be used to augment their nutrition and control disease. It has been remarked by many that the number of mammalian species which have been domesticated is extremely small relative to the known number of species, and that the continent of Africa, so rich in herbivorous mammals, has not contributed one species to the list of domesticated stock. Several attempts at new domestications are currently in progress, notably with the common eland, which, as evident in Table 16.1, could be a contender. Russian studies which have been in progress since the 1890s show that these animals can be tamed and made amenable (Treus & Kravchenko, 1968). The increase in the population is currently 12 to 14 % per annum, a figure which is relevant to the discussion of cropping rate given for wild stocks. Posselt (1963) established a domesticated herd at Lupani in Southern Rhodesia in 1954, and has described its history, including that of a subsidiary herd of twenty-four, reduced in six months to eleven by predators (leopards) and poachers. The original herd was transferred to the Department of Parks in Rhodesia and investigations continue (Kerr *et al.*, 1970; Kerr & Roth, 1970; Roth, 1970; Roth & Osterberg, 1971). Additional studies are being made with springbok (Skinner *et al.*, 1971) largely with a view to game ranching of a stock to which an ownership can be assigned rather than to domestication. In temperate areas, partial domestication of the red deer is being studied in Scotland under an IBP programme (Bannerman & Blaxter, 1969; Blaxter, 1972).

In conclusion, there are many alternative strategies to be considered in producing meat from wild species. Meat production could arise from the selective culling of wild species kept in reservations, or the cropping of a wild stock with concomitant control of predators. It might involve a combination of trophy hunting with meat production, or the co-existence of wild and domestic species in an extensive type of ranching. It could equally well involve a containment and agricultural manipulation of a species as implied by terms such as domestication or wild-life farming. It also seems realistic to consider pasturing game on good farm land or even fattening such animals indoors where the species is sufficiently manageable, and, in an established market, game meat is priced favourably relative to meat from conventional stock. All approaches involve complex ecological and technical problems.

References

Asibey, E. O. A. (1966). Why not bushmeat too? *Ghana Farmer*, **10**, 165–70.

Bannerman, M. M. & Blaxter, K. L. (1969). *The husbanding of red deer.* Highlands & Islands Development Board and Rowett Research Institute, Aberdeen.

Bland, G. H. M. (1969). The potential market for venison and its possible operation. In: *The husbanding of red deer* (ed. M. M. Bannerman & K. L. Blaxter), pp. 13–16. Highlands & Islands Development Board and Rowett Research Institute, Aberdeen.

Blaxter, K. L. (1972). Deer farming. *Scottish Agriculture*, **51**, 225–30.
Crawford, M. A. (1968*a*). Food selection under natural conditions and the possible relation to heart disease in man. *Proceedings of the Nutrition Society*, **27**, 163–71.
Crawford, M. A. (1968*b*). Fatty-acid ratios in free living and domestic animals. Possible implications for atheroma. *Lancet*, **i**, 1329–33.
Crawford, M. A. (1968*c*). Possible use of wild animals as future sources of food in Africa. *Veterinary Record*, **82**, 305–18.
Crawford, M. A., Gale, M. M., Woodford, M. H. & Casped, N. M. (1970). Comparative studies on fatty acid composition of wild and domesticated meats. *International Journal of Biochemistry*, **1**, 295–305.
Darling, F. F. (1960). Wildlife in an African territory. *Scientific American*, **203**, 123–8.
Devendra, C. & Burns, M. (1970). *Goat production in the tropics*. Technical Communication No. 19. Commonwealth Bureau of Animal Breeding & Genetics, CAB Farnham Royal, England.
De Vos, A. (1969). Ecological conditions affecting the production of wild herbivorous mammals on grassland. *Advances in Ecological Research*, **6**, 137–83.
Garton, G. A. (1969). Polyunsaturated fatty acids in ruminant tissues in relation to atheroma in man. *Lancet*, **i**, 1217–18.
Golley, F. B. & Buechner, H. K. (1968). *A practical guide to the study of the productivity of large herbivores*. Blackwell Scientific, Oxford.
Halloran, A. F. (1957). Live and dressed weights of American bison. *Journal of Mammalogy*, **38**, 139–40.
Harthoorn, A. (1968). Cropping of wild herbivores. *World Review of Animal Production*, **4**, 120–7.
Hopcraft, D. (1970). East Africa: the advantages of farming game. *Span*, **13**, 29–32.
Kay, R. N. B. (1970). Meat production from wild herbivores. *Proceedings of the Nutrition Society*, **29**, 271–8.
Kerr, M. A. & Roth, H. H. (1970). Studies on the agricultural utilization of semi-domesticated eland. 3. Horn development and tooth eruption as indicators of age. *Rhodesian Journal of Agricultural Research*, **8**, 149–56.
Kerr, M. A., Wilson, V. J. & Roth, H. H. (1970). Studies on the agricultural utilization of semi-domesticated eland. 2. Feeding habits and food preferences. *Rhodesian Journal of Agricultural Research*, **8**, 71–7.
Kyle, R. (1972). *Meat production in Africa*. University of Bristol.
Ledger, H. P. (1963). A note on the relative body composition of wild and domesticated ruminants. *Bulletin of Epizootic Diseases of Africa*, **11**, 163–5.
Ledger, H. P. (1968). Body composition as a basis for a comparative study of some East African mammals. *Symposia of the Zoological Society of London*, **21**, 289–310.
Mann, I. (1964). Vitamin content and amino acid composition of some African game animals. *Journal of Agricultural and Food Chemistry*, **12**, 374–6.

Mitchell, B. (1972). Annual cycle of condition and body composition of red deer on the island of Rhum. *Deer, Journal of the British Deer Society*, **2**, 904–7.

Mitchell, G. J. (1971). Measurements, weights and carcass yields of pronghorns in Alberta. *Journal of Wildlife Management*, **35**, 76–85.

Posselt, J. (1963). The domestication of the eland. *Rhodesian Journal of Agricultural Research*, **1**, 81–7.

Riney, T. (1964). The economical use of wildlife in terms of its productivity and development as an agricultural activity. FAO Regional Meeting, Addis Ababa, paper No. 49, March 1964.

Roth, H. H. (1970). Studies on the agricultural utilization of semi-domesticated eland. 1. Introduction. *Rhodesian Journal of Agricultural Research*, **8**, 67–70.

Roth, H. H. & Osterberg, R. (1971). Studies on the agricultural utilization of semi-domesticated eland. 4. Chemical composition of eland browse. *Rhodesian Journal of Agricultural Research*, **9**, 45–52.

Skinner, J. D. (1970). Game ranching in Africa as a source of meat for local consumption and export. *Tropical Animal Health and Production*, **2**, 151–7.

Skinner, J. D., von la Chevallerie, M. & van Zyl, J. H. M. (1971). An appraisal of the springbok for diversifying animal production in Africa. *Animal Breeding Abstracts*, **39**, 215–24.

Talbot, L. M. (1963). Comparison of the efficiency of wild animals and domestic species in utilization of East African Rangelands. *Publications of the IUCN NS*, **1**, 328–35.

Talbot, L. M., Payne, W. J. A., Ledger, H. P., Verdcourt, L. D. & Talbot, M. H. (1965). *The meat production potential of wild animals in Africa. A review of biological knowledge.* Technical Communication No. 16 of the Commonwealth Bureau of Animal Breeding & Genetics, Edinburgh.

Treus, V. & Kravchenko, D. (1968). Methods of rearing and economic utilization of eland in the Askaniya Nova Zoological Park. *Symposia of the Zoological Society of London*, **21**, 395–411.

van Zyl, J. H. M., von la Chevallerie, M. & Skinner, J. D. (1969). A note on the dressing percentage in the springbok and impala. *Proceedings of the South African Society of Animal Production*, **8**, 199–200.

Venjarskij, A. D. (1963). Effectivnost' nagula lošadej. *Trudy Semipalatinsk Zoovet. Inst.*, **3**, 87.

Venjarskij, D. (1959). O mjasnom nagule lošadej. *Konevodstvo*, **7**, 16–17.

von la Chevallerie, M. (1970). Meat production from wild ungulates. *Proceedings of the South African Society of Animal Production*, **9**, 73–87.

von la Chevallerie, M. (1972). Meat quality of seven wild ungulate species. *South African Journal of Animal Science*, **2**, 101–3.

Walters, C. J. & Bandy, P. J. (1972). Periodic harvest as a method of increasing big game yields. *Journal of Wildlife Management*, **36**, 128–34.

Young, E. & van den Heever, L. W. (1969). African buffalo as a source of food and by-products. *Journal of the South African Veterinary Medical Association*, **40**, 83–8.

Young, E. & Wagener, L. J. J. (1968). The impala as a source of food and by-products. *Journal of the South African Veterinary Medical Association*, **39**, 81–6.

Young, E., Wagener, L. J. J. & Bronkhorst, P. J. L. (1969). The production potential, parasites and pathology of free living wildebeest of the Kruger National Park. *Journal of the South African Veterinary Medical Association*, **40**, 315–18.

17. The use of non-protein nitrogen by ruminants

T. R. PRESTON

Whenever there is discussion at international level about the means of bridging the protein gap, it has been customary to conclude that the solution to the problem must be through the growing of agricultural crops as sources of vegetable protein for direct human consumption. It has been argued that domestic animals are inefficient producers of protein, and that their role can never be other than as a source of luxury proteins for consumption by the elite fraction of the population that can afford to pay the high prices required for such products.

It is true that in many parts of the world, particularly the temperate regions, protein is produced most efficiently by growing grains and pulses for direct consumption, or by extracting protein from high-yielding grasses and legumes. However, these procedures are not necessarily the most appropriate for tropical regions, particularly the humid tropics, for these areas are better adapted to producing high carbohydrate crops such as sugar cane and cassava. Such crops have a negligible content of protein, but they are excellent substrates for the growth of micro-organisms, which can produce high quality protein in the form of their own protoplasm, using as nitrogen (N) sources simple chemicals such as ammonia and urea. Several proposals have been put forward for this type of industrial protein production; however, most suffer from the low palatability of the final product and also certain health hazards, according to the nature of the substrate used for the fermentation.

Fermentation can also take place within the digestive tract of animals, and ruminants are particularly appropriate for this since the microbial protein formed by fermentation is digested subsequently by gastric enzymes, the net result being an overall transformation of simple N compounds in the feed into tissue and milk proteins of both high biological value and palatability. In this respect, ruminants possess singular advantages over non-ruminant animals such as pigs and poultry which do not have this facility. Their role is no more than that of modifying the biological value of low quality feed protein, and this is achieved only at considerable cost since the efficiency of the conversion process, on a protein basis, is at best some 18 % for egg production and at worst 12 % as in the case of pigmeat (Preston, 1971).

The hypothesis then is that ruminants, by virtue of their symbiotic microflora, bring about a net synthesis of high quality edible protein from chemical N and fermentable carbohydrate. The practical demonstration of this concept can be seen in Cuba, where half a million cattle are fattened each year on diets in which 80 % of the energy is in the form of molasses and over 60 % of the N is urea (Preston, 1972; A. Molina, personal communication).

The economics of this process depend on the efficiency with which the carbohydrate source serves as an energy substrate for microbial growth. Some practical estimates of this process can be calculated from results provided by Preston (1972) from an experiment where a diet of fermentable carbohydrate (principally molasses) and non-protein N was fed to fattening cattle. In this trial the conversion into liveweight of the non-protein diet was 95 MJ metabolisable energy (ME) as carbohydrate for each kilogram of liveweight gain. Other data, from trials with similar animals and diets (Willis *et al.*, 1968) indicate that, on average, 1 kg of liveweight gain contains 100 g of dry protein; therefore, the final conversion is 950 MJ ME as carbohydrate for 1 kg of dry protein in the form of meat.

To convert these data into estimates of protein production per hectare it is proposed to take as an example the sugar cane crop, which can be considered as the highest yielding crop which is grown in the humid tropics on a commercial scale. Although this crop so far has been grown exclusively for sugar production, recent developments (Miller, 1973) indicate that when the indigestible rind is removed from the cane stalk, the remaining part of the plant is an excellent carbohydrate source for ruminant feeding (Donefer, 1973). Experiments in Barbados (Warnaars, 1973) have shown that a sugar cane crop with a potential yield of 100 t (tonnes) per hectare,† after derinding, produces the equivalent of 34.3 t of feed dry matter (DM). This feed has been shown to have a digestibility of 70 % (see Pigden, 1973) and, assuming that 1 kg of digestible DM is equivalent to 15 MJ of ME, then the potential production is some 360000 MJ ME/ha. Using the conversion figure of 950 MJ ME/kg dry meat protein, then the protein production per hectare is 378 kg.

† The average yield of sugar cane on a world basis is some 50 t/ha; 100 t/ha is what can be achieved fairly easily with a reasonable degree of technology, as for example in Peru and Ethiopia (FAO, 1971).

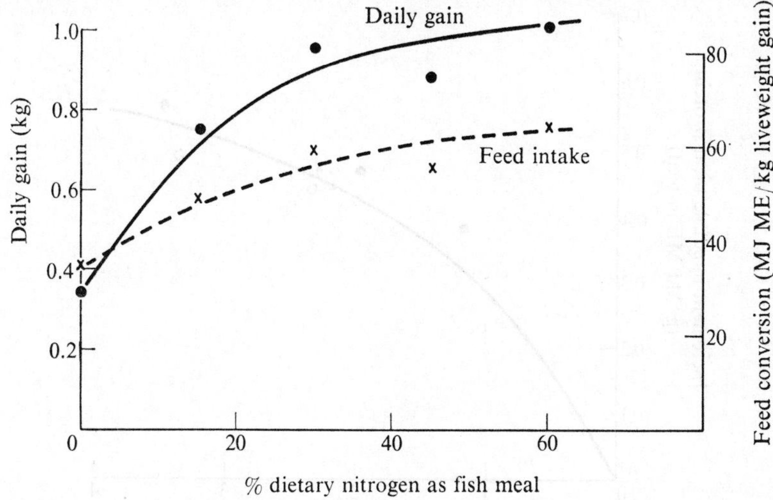

Fig. 17.1. Effect of replacing urea nitrogen with true protein nitrogen (fish meal) on feed intake and daily gain in fattening cattle fed diets based on molasses and urea (from Preston, 1972).

The efficiency of the ruminant animal in converting fermentable carbohydrate into tissue or milk protein is increased if some feed protein is included in its diet. This is because cattle eat a diet containing no true protein less readily than one containing some true protein (see Fig. 17.1), and concomitant with increased intake there is also more rapid growth and better overall efficiency of feed conversion in terms of energy. Furthermore, we observe that the addition of 6 g of feed protein per 4.2 MJ ME increases protein yield from 378 to 508 kg/ha. For the 360000 MJ considered in the last paragraph, 522 kg of true protein supplement would be needed. The effect can be considered in two ways: if attention is confined to true protein, 522 kg in the feed yields 508 kg in meat i.e., a net efficiency of 97 %; but considering marginal efficiency only, 522 kg has produced an extra 130 kg of protein in meat i.e., 25 % (see Fig. 17.2). Still further increments of feed protein, contribute to increased efficiency of conversion of energy into protein, but the marginal efficiency of converting feed protein into meat protein diminishes steadily following the general law of diminishing returns.

It is not yet possible to draw up similar relationships for conversion of urea into milk protein. Virtanen (1966) has published results of an experiment with a limited number of cows receiving a semi-synthetic diet in which all the N was in non-protein form. His findings have yet

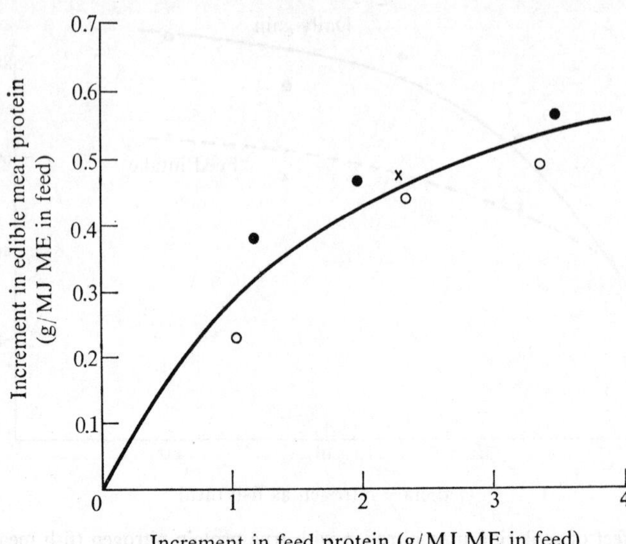

Fig. 17.2. Marginal efficiency of increasing meat protein output from urea/molasses diets given to beef cattle when supplementary true protein is added to the diet (from Preston, 1972).

to be confirmed on a commercial scale, but from theoretical considerations, similar relationships could be expected between crop yield of fermentable carbohydrate and milk protein output.

The above discussion has centred on the use of urea as the source of non-protein N. The logic for this is that, while many inorganic and organic compounds give rise to ammonia on hydrolysis, and thus can serve as substrates for microbial growth, only urea has so far been used on a widespread commercial scale. Of the other chemicals, biuret, which is a condensation product of urea, has been subjected to most experimentation. Its main advantage is its slower release of ammonia (compared with urea) and hence a reduced risk of ammonia toxicity. Its disadvantage is that the bacteria required to hydrolyse biuret apparently are not abundant in normal rumen liquor, and a period of adaptation is required whenever animals are introduced to a biuret-containing ration (Johnson & Clemens, 1973). Furthermore, the slow release of ammonia from biuret can be a disadvantage in diets with large amounts of readily fermentable carbohydrates; ammonia is then released too slowly to supply the N needs of the rapidly growing micro-organisms. Thus, Gonzalez & Harcus (unpublished data) found that in fattening bulls fed a basal diet of unlimited molasses, restricted forage and fish meal, the growth rate was some 50 % faster when the supplementary source of N was urea rather than biuret.

The same constraints seem to apply to N from poultry excreta. In a molasses-based diet, with small amounts of fish meal and forage, growth rate decreased when excreta from laying cages was used to replace urea as the major source of supplementary N (approximately 60 % of total diet N) (Preston *et al.*, 1970). The determining factor appears to be the proportion of diet N supplied by the excreta. Thus, in several experiments, poultry waste, either from laying cages or as litter from broiler operations, at levels of up to 40 % of the diet, was usually as good as oilseed meals; but in all these cases the remainder of the ration was cereal grain, usually maize (El-Sabban *et al.*, 1970; Fontenot *et al.*, 1970; Galmez *et al.*, 1970; Cuesta *et al.*, 1972; Cullinson *et al.*, 1973). In such a situation, at least 60 % of the N ration is preformed protein from the cereal grain, whereas when molasses or sugar cane is the basal energy source, negligible amounts of preformed protein are supplied.

The above conclusions apply in the main to poultry droppings used without prior fermentation, other than the small amount that may take place in deep litter wastes. Present interest is in more refined methods of recycling wastes from both ruminants and non-ruminants, in which aerobic fermentation is encouraged with the aim of eventually harvesting the single-cell protein resulting from microbial growth. The liquid phase resulting from an oxidation ditch used to process both solid and liquid excreta from a pig fattening operation had 3 % dry matter, and, in the dry matter, 49 % protein, 1.42 % lysine, and all other essential as well as non-essential amino acids. When this fluid was added to a pig ration marginally deficient in protein and in lysine, growth rate and feed efficiency were improved by some 6 to 7 % (Harmon *et al.*, 1973).

This latter approach has good potentialities because it is a means of achieving a net synthesis of protein from simple N compounds in the animal waste; in contrast, direct use of raw animal waste in ruminant rations is only economically feasible when the remainder of the diet contains substantial amounts (more than 60 % of requirements) of preformed protein. Use of animal waste for non-ruminant feeding can rarely be recommended since performance is depressed above about 10 % replacement (Flegal & Dorn, 1971). Thus, for both ruminant and non-ruminant feeding, unprocessed animal waste is perhaps most appropriately assessed in terms of its energy value. According to Polin *et al.* (1971), the energy contribution (5.5 MJ ME/kg on 90 % DM basis) of poultry waste (from laying birds) was what would be expected from a typical fibrous feed such as dried alfalfa or wheat bran.

References

Cuesta, P. M., Martinez, M. C., Serrano, J. T. & Gutiérrez, A. J. (1972). Urea y gallinaza en alimentación de corderos. (Urea and poultry litter in the feeding of lambs.) In: Second World Congress of Animal Feeding, Madrid.

Cullinson, A. E., McCampbell, H. C. & Warren, E. P. (1973). Use of dried broiler feces in steer rations. *Journal of Animal Science*, **36**, 218.

Donefer, W. (1973). Comfith as an animal feed. In: CIDA seminar on sugar cane as livestock feed, Barbados. Canadian International Development Agency, Ottawa.

El-Sabban, F. F., Bratzler, J. W., Long, T. A., Frear, D. E. H. & Gentry, R. F. (1970). Value of processed poultry waste as a feed for ruminants. *Journal of Animal Science*, **31**, 1.

FAO (1971). *Production Yearbook*, 1970–1. FAO, Rome.

Flegal, C. J. & Dorn, D. A. (1971). The effects of continually recycling dehydrated poultry waste (DPW) on the performance of SCWL laying hens; a preliminary report. Research Report No. 152, Farm Science, Michigan Agricultural Experiment Station, East Lansing, Michigan.

Fontenot, J. P., Tucker, R. W., Harmon, B. W., Libke, K. G. & Moore, W. E. C. (1970). Effects of feeding different levels of broiler litter to sheep. *Journal of Animal Science*, **30**, 2.

Galmez, J., Santisteban, E., Haardt, E., Crempien, C., Villalta, L. & Torrel, D. (1970). Performance of ewes and lambs fed broiler litter. *Journal of Animal Science*, **31**, 1.

Harmon, B. G., Day, D. L., Baker, D. H. & Jensen, A. H. (1973). Nutritive value of aerobically or anaerobically processed swine waste. *Journal of Animal Science*, **37**, 510.

Johnson, R. R. & Clemens, E. T. (1973). Adaptation of rumen micro-organisms to biuret as an NPN-source to low quality roughage rations. *Journal of Nutrition*, **103**, 494.

Miller, R. B. (1973). Theory and practice of sugar cane separation. In: CIDA seminar on sugar cane as livestock feed, Barbados. Canadian International Development Board, Ottawa.

Pigden, W. J. (1973). Evaluation of comfith as a commercial livestock feed in the Caribbean. In: CIDA seminar on sugar cane as livestock feed, Barbados. Canadian International Development Board, Ottawa.

Polin, D., Verghese, S., Neff, M., Gomez, M., Flegal, C. J. & Zindel, H. (1971). The metabolizable energy value of dried poultry waste. Research Report No. 152, Farm Science, Michigan Agricultural Experiment Station, East Lansing, Michigan.

Preston, T. R. (1971). The use of urea in high molasses diets for milk and beef production in the humid tropics. Report of *ad hoc* consultation on the value of non-protein nitrogen for ruminants consuming poor herbages. FAO, Rome.

Preston, T. R. (1972). Molasses as an energy source for cattle. *World Review of Nutrition and Dietetics*, **17**, 250.

Preston, T. R., Willis, M. B. & Elias, A. (1970). The performance of two breeds given different amounts and sources of protein in a high molasses diet. *Animal Production*, **12**, 457.

Virtanen, A. I. (1966). Milk production of cows on a protein-free feed. *Science*, **153**, 1603.

Warnaars, B. C. (1973). Growing of sugar cane as an animal feed. In: CIDA seminar on sugar cane as livestock feed, Barbados. Canadian International Development Agency, Ottawa.

Willis, M. B., Preston, T. R., Martin, J. L. & Velazquez, M. (1968). Carcass composition of Brahman bulls fed high energy diets and slaughtered at different live weights. *Revista cubana de Ciencia agricola*, **2**, 87. (English edition.)

18. Non-protein nitrogen in pig nutrition

R. BRAUDE

In the past, in order to express the dietary contribution of nitrogenous components, their total nitrogen content (N) was estimated and a simple term of crude protein (CP) was created (N × 6.25 or an adjusted coefficient). This was widely applied in connection with animal requirements and composition of feedstuffs. The term CP served both the science and the practice of feeding well, but for a long time now its inadequacy has been recognised.

There is a general agreement about the dietary importance of non-protein nitrogen (NPN) given to monogastric animals in the form of free essential amino acids (AA), but conflict arises over the extent to which non-essential AA and the other N-compounds included in CP can be used.

Recently, Eggum & Christensen (1973) comprehensively reviewed the subject of NPN in the nutrition of monogastric animals. As far as pigs are concerned, their evidence can be augmented by reference to some of the earlier papers: Abderhalden & Lampe (1913), Grafe (1913), Piepenbrock (1927), Braude & Foot (1942) and Shelton *et al.* (1950). All these dealt with feeding urea to pigs and produced rather confusing results, but the fact that some found NPN useful kept the issue open. The controversy continued, and within the last decade Burkser *et al.* (1965), Kornegay *et al.* (1965, 1970), Meacham & Thomas (1966–7), Baia *et al.* (1967) and Kornegay (1972) claimed that in some circumstances urea was utilised by young growing pigs, while Bowland (1967), Meacham *et al.* (1967–8), Pastuszewska (1967), Hintz *et al.* (1969), Grimson & Bowland (1971) and Tylecek *et al.* (1971) presented contrary evidence.

In recent years attention moved to sources of NPN other than urea. Barber *et al.* (1969) attempted unsuccessfully to improve a cereal diet supplemented with lysine by an addition of diammonium phosphate (DAP). Velasquez *et al.* (1970) using DAP and Wehrbein *et al.* (1970) using DAP and ammonium citrate (DAC) found these supplements nutritionally inert. On the other hand Garanina & Kosharov (1972) recorded positive responses to DAP and DAC.

Efforts are continuing to resolve the controversy on whether NPN can be utilised by pigs and the scope of the recent investigations has widened.

Akulinin & Cingovatov (1967) fitted gastric fistulae in two pigs and reported that when urea was added (equivalent to one-fifth of CP content) the proteolytic activity of stomach contents was reduced. The weight gains were also reduced. Aliev & Kosharov (1971) placed a catheter in the portal veins of two pigs and infused ^{15}N-labelled glycine and ^{15}N-labelled urea directly by cannula into the caecum and concluded that N from glycine, urea and ammonium salts was used for the synthesis of AA and proteins. Grimson *et al.* (1971) using ^{15}N-labelled urea also indicated that dietary urea is used in protein synthesis by the pig. Flam & Bednarova (1971) concluded that the utilisation of a part of the N of synthetic urea by the bacteria in the pig digestive tract was possible. Rerat & Aumaitre (1971) fed a well-balanced meal with and without urea to young pigs and studied the variation in the level of urea and ammonia in the portal and peripheral blood in order to follow the transit through the digestive tract, reabsorption by the intestines and passage through the liver.

Recently, workers from the Guy's Hospital, London and the Rowett Research Institute, Aberdeen discussed purine metabolism in pigs in a series of papers (Simmonds *et al.*, 1973*a*, *b*; Cameron *et al.*, 1973) which provided information on relevant enzyme activities and metabolic pathways which may eventually help in solving the problem of utilisation of NPN compounds by the pig.

Our need for understanding the role of NPN has become more urgent. At a time when great anxiety is voiced about the future supplies of protein for man and animals, it may be opportune to renew efforts to explore whether a case can be made for effective supplementation of diets with relatively inexpensive simple nitrogenous compounds.

The extent of our interest in NPN has been widening, and two examples from the current research programme of my own department illustrate this clearly.

In the search for sources of protein for pigs we became interested in the yeast and bacterial proteins produced industrially. Amongst others, we have tested the BP-yeast (Barber *et al.*, 1971) and the ICI-protein (not yet reported) and found them both nutritionally adequate for growing pigs. However, the BP-yeast contains about 8–10 %, and the ICI-protein about 14–16 % (it may be as high as 24 %) of nucleic acids, and we have to admit that our knowledge of the utilisation of these nitrogenous compounds by the pig is virtually nil. Claims have been made that high nucleic acids in single-cell proteins make them unacceptable for human diets and that different species of monogastric animals may react differently to relatively high intakes of nucleic acids (cf. Worgan, 1973). In fact, no guide exists on the nutritive value of nucleic acids for pigs or on the safety of including a high level in the diet.

The second example refers to the problem which arose when we became interested in the extraction of juice from lucerne and studied its nutritional value for pigs (Barber *et al.*, 1973). We have established that rapid changes in the nitrogenous components of the juice occur very soon after mechanical removal of the juice from the plant. Enzymic and bacterial action appear to be involved. It has certainly become clear that from the nutritional point of view the terms total N or CP have great limitations, and that in order to make an adequate assessment an understanding of the nature and nutritional contribution, if any, of the NPN fraction becomes essential. The NPN is obviously a mixture of compounds. Part of it could be amino acid N or peptides, and thus be generally useful; part could be nucleic acids and their usefulness may depend on circumstances; another part could simply be useless. In fact, once again we had to admit that, at present, we do not possess enough knowledge to resolve this problem.

References

Abderhalden, E. & Lampe, A. E. (1913). *Zeitschrift für Physiologische Chemie*, **84**, 218.

Akulinin, A. & Cingovatov, V. A. (1967). *Nauchni Trudy Omsk. Vet. Inst.*, **24**, 109.

Aliev, A. A. & Kosharov, A. N. (1971). *Vest. sel'skokhoz. Nauki, Mosk.*, **16**, 27.

Baia, G., Arisanu, I., Vintila, M. & Gheorghiu, V. (1967). *Lucr. stiint. Inst. Cerc. zooteh.*, **25**, 597.

Barber, R. S., Braude, R., Florence, E., Mitchell, K. G. & Newport, M. J. (1973). *Proceedings of the British Society of Animal Production*, **2**, 89.

Barber, R. S., Braude, R. & Mitchell, K. G. (1969). *Animal Production*, **11**, 292.

Barber, R. S., Braude, R., Mitchell, K. G. & Myres, A. W. (1971). *British Journal of Nutrition*, **25**, 285.

Bowland, J. P. (1967). 46th Annual Feeders Day, Department of Animal Science, University of Alberta, p. 20.

Braude, R. & Foot, A. S. (1942). *Journal of Agricultural Science, Cambridge*, **32**, 70.

Burkser, G. V., Tihinova, M. B. & Ibraev, K. I. (1965). *Svinovodstvo*, **19**, 28.

Cameron, J. S., Simmonds, H. A., Hatfield, P. J., Jones, A. S. & Cadenhead, A. (1973). *Israel Journal of Medical Sciences*, **9**, 1087.

Eggum, B. O. & Christensen, K. D. (1973). *Zeitschrift für Tierphysiologie, Tierernährung und Futtermittelkunde*, **31**, 332.

Flam, F. & Bednarova, M. (1971). *Zivocisna Vyroba*, **16**, 889.

Garanina, N. A. & Kosharov, A. N. (1972). *Dokl. vses. Akad. sel'skokhoz. Nauk.*, (11), 33.

Grafe, E. (1913). *Zeitschrift für Physiologische Chemie*, **84**, 69.

Grimson, R. E. & Bowland, J. P. (1971). *Journal of Animal Science*, **33**, 58.

Grimson, R. E., Bowland, J. P. & Milligan, L. P. (1971). *Canadian Journal of Animal Science*, **51**, 103.

Hintz, H. F., Pond, W. G. & Visek, W. J. (1969). *Animal Production*, **11**, 553.

Kornegay, E. T. (1972). *Journal of Animal Science*, **34**, 55.

Kornegay, E. T., Miller, E. R., Ullrey, D. E., Vincent, B. H. & Hoefer, J. A. (1965). *Journal of Animal Science*, **24**, 951.

Kornegay, E. T., Mosanghini, V. & Snee, H. D. (1970). *Journal of Nutrition*, **100**, 330.

Meacham, T. N. & Thomas, H. R. (1966–7). Virginia Polytechnic Institute Research Division, Livestock Report No. 122, p. 36.

Meacham, T. N., Thomas, H. R. & Horsley, W. E., Jr (1967–8). Virginia Polytechnic Institute Research Division, Livestock Report No. 126, p. 82.

Pastuszewska, B. (1967). *Rocznik nauk Rolniczych*, Ser. B, **89**, 503.

Piepenbrock, A. (1927). *Fortschr. Landw.*, **2**, 650.

Rerat, A. & Aumaitre, A. (1971). *Annales de Biologie animale, Biochimie et Biophysique*, **11**, 348.

Shelton, D. C., Beeson, W. M. & Mertz, E. T. (1950). *Archives of Biochemistry and Biophysics*, **29**, 446.

Simmonds, H. A., Hatfield, P. J., Cameron, J. S., Jones, A. S. & Cadenhead, A. (1973*a*). *Biochemical Pharmacology*, **22**, 2537.

Simmonds, H. A., Rising, T. J., Cadenhead, A., Hatfield, P. J., Jones, A. S. & Cameron, J. S. (1973*b*). *Biochemical Pharmacology*, **22**, 2553.

Tylecek, J., Skalova, J. & Zednik, M. (1971). *Biologizace Chem. Vyz. Zvir.*, **7**, 165.

Velasquez, M., Preston, T. R. & Macleod, N. A. (1970). *Revista Cubana de Ciencia Agricola* (English edition) **4**, 105.

Wehrbein, G. F., Vipperman, P. E., Jr, Peo, E. R., Jr & Cunningham, P. J. (1970). *Journal of Animal Science*, **31**, 327.

Worgan, J. T. (1973). In: *The biological efficiency of protein production* (ed. J. G. W. Jones), p. 339. Cambridge University Press, London.

19. The domestic non-ruminant animal as consumer and provider of protein

A. A. WOODHAM

Pigs and poultry are the principal non-ruminant domestic animals; both contribute largely at present to world stocks of edible protein. From Table 19.1 it can be seen that over the last five years they contributed almost one-third of the total animal protein available for food. Unfortunately they also consume considerable amounts of high quality protein which could be consumed directly by man with greater efficiency. Precise calculations of the efficiency with which plant proteins are converted into animal proteins by passage through non-ruminants are difficult to make, but it has been estimated that for the hen it is about 25–30 % for egg production and 20–25 % for meat production, and for the pig, 10–20 % (Halnan, 1941; Boyd, 1951; Leitch & Godden, 1953). This appears to represent a wasteful use of grain and concentrate protein, but it should be borne in mind that egg and muscle protein are of higher nutritive value than grain protein as well as of many of the plant concentrates which could be fed. Furthermore the efficiency of the non-ruminant compares very favourably with the ruminant as a meat producer; that of beef cattle ranging from 5 to 10 % and sheep from 10 to 13 % according to the sources already quoted. More recent calculations have not materially affected the picture. Byerly (1967) concludes that under optimum conditions young individuals of each livestock species may convert about one-third of the digestible dietary protein into tissue protein. Approximately one-half of this may be suitable for food and hence the overall efficiency is around 15 %. Laying hens and lactating mammals may convert up to 30 and 50 % of dietary protein into food protein respectively. Overall it requires about 7.5 kg of digestible protein to produce 1 kg of food protein in the form of meat, milk or eggs. Lodge (1970) has confirmed this ratio for cattle, sheep and pigs. Older cattle and sheep are less efficient; they need 13.0 kg and 10.0 kg of dietary protein to produce 1 kg of muscle protein. The young pig is particularly efficient; only 6.5 kg of dietary protein are necessary to produce 1 kg of muscle protein in a 5-month-old animal. If the dam's food requirements are taken into account the superiority of the young pig over the ruminant is even more marked. Because the ratio of progeny to dam is so much greater than that for ruminants, the additional feed required by the latter has an almost negligible effect on the overall protein conversion. In fact Lodge quotes a figure of 8 kg dietary protein needed to produce 1 kg of muscle protein for both 5- and 7-month-old pigs. Calculations using data from world wide sources (Panel on the World Food Supply, 1967) agree on a figure of 8.0 for pork production but arrive at somewhat better efficiencies for poultry – 4.3 for egg production and 5.5 for poultry meat.

Conditions of increasing protein stringency, allied to the need to make optimal use of land resources, will tend to favour the extension of pig and poultry raising in the short term due to their suitability for intensive farming practices. One must envisage, therefore, that these animals will be of increasing importance so long as it is possible for mankind to consume animal products at all. The efficiency figures quoted above are based only on the conversion of feed protein. In terms of land utilisation the efficiency of the intensively housed non-ruminant and especially of the broiler chicken is still greater than that of the ruminant.

Non-ruminants need high quality protein because of their inability to synthesise certain amino acids (described in consequence as 'essential'). Feedstuffs providing comparatively high levels of these amino acids, particularly lysine and methionine, are in demand for human feeding so that the non-ruminant competes with man for these. A wide variety of plant proteins are regularly included in non-ruminant diets. These differ from one another not only in their amino acid make-up but also in the extent to which they are associated with toxins, growth inhibitors and other antimetabolites, and in the way in which they are affected by processing. The factors which affect the nutritive value for non-ruminants of even such traditional feedstuffs as soybean (*Glycine max*) and groundnut meals (*Arachis hypogaea*) are not yet fully understood. However, commercially produced oilseed meals from these more important types are generally found to be consistent in their feeding value, suggesting that, probably as a result of simple trial and error, satisfactory processing conditions have been achieved and can be maintained within fairly close limits of temperature, time, moisture, etc. Problems remain with regard to the less common oilseeds and particularly the potentially important members of the Cruciferae. Though possessing a good amino acid composition compared to many of the commonly used plant proteins, they require special processing in order

Table 19.1. *World production of protein from livestock 1966–71 (megatonnes)*

Year	Beef, veal and buffalo meat		Pork		Poultrymeat		Hen eggs		Milk		Total protein from all livestock	% contribution by:			% contribution by non-ruminants
	Production	Edible protein	Production	Edible protein	Production	Edible protein	Production	Edible protein	Production	Edible protein		Pork	Poultry-meat	Eggs	
1966	34.8	5.74	31.1	3.17	13.9	2.59	17.7	2.28	379.9	13.3	27.08	11.7	9.56	8.42	29.7
1967	36.1	5.96	32.3	3.29	14.7	2.74	18.9	2.44	387.0	13.5	27.93	11.8	9.81	8.74	30.4
1968	37.8	6.24	32.9	3.36	15.0	2.79	19.4	2.50	395.0	13.8	28.69	11.7	9.72	8.71	30.1
1969	38.6	6.37	32.9	3.36	15.9	2.96	20.1	2.59	396.9	13.9	29.18	11.5	10.14	8.88	30.5
1970	39.0	6.44	34.3	3.50	17.4	3.24	21.0	2.71	398.5	13.9	29.79	11.7	10.88	9.10	31.7
1971	38.9	6.42	35.6	3.63	NI		21.6	2.79	400.9	14.0	—	—	—	—	—

NI: no information available.
Edible protein conversion factors: beef, veal and buffalo, 16.5; pork, 10.2; poultry, 18.6; eggs, 12.9; milk, 3.5 (Panel on the World Food Supply, 1967).
Data from FAO (1966–71).

to remove toxic principles which at present limit the possible levels of inclusion in non-ruminant diets. Considerable progress has been made in recent years, however, and it is interesting that early drastic detoxification procedures aimed at destroying or inactivating the toxins without considering the possible effects upon the protein itself, have given way to more gentle procedures which are nevertheless claimed to be equally efficient. Inactivation of rapeseed (*Brassica napus*) toxins, for example, by dry heating to 130 °C (Frölich, 1953) satisfactorily destroyed the toxic principle but also undoubtedly severely damaged the protein. Ammoniation of crambe (*Crambe abyssinica*) seed was similarly effective with respect to the toxin (Kirk *et al.*, 1966) but the palatability was adversely affected and an alternative procedure was eventually recommended (Mustakas *et al.*, 1968). Detoxification is also necessary in the case of cottonseed meal (*Gossypium herbaceum*) where the gossypol content can limit the amount which can be incorporated into non-ruminant diets. As with the Cruciferae, the bulk of the work has been concentrated on solvent-extraction techniques and satisfactory products have been achieved using the right combinations of solvents (Mann *et al.*, 1962; Pons & Eaves, 1967). It is frequently stated that animals may act as 'toxin filters', being able to utilise food that could not be consumed directly by man and yielding non-poisonous products. Close examination of such claims generally reveals that the animal, and especially the non-ruminant, thrives best on high quality foods and the economic penalty exacted by the use of inferior feedstuffs is in practice a serious deterrent. Since the aflatoxin disaster no-one would countenance the feeding of moulding grain, though this was at one time thought to be satisfactory for some classes of animals. Most plant toxins depress growth or produce damage in vital organs, though there may be no effect upon the meat produced.

Improvements in processing techniques are an obvious way to ensure that the maximum value is achieved from an existing range of protein concentrates. When the ideal conditions have been achieved, however, further advance depends upon the production of superior raw materials. Much effort has been devoted by plant breeders in recent years to the development of strains of cereals and oilseeds which have more desirable agronomic properties. These include improved disease-resistance, higher crop yields and better harvesting characteristics. Only recently has attention been given to the question of improving the quality of the protein. With animals as discriminating as non-ruminants in their nutritional requirements this matter is clearly of considerable import-

Table 19.2. *Requirements of the chick and of the growing pig (up to 50 kg liveweight) for amino acids (% of diet)*

	Pig	Chick
Threonine	0.5	0.65
Glycine	—	0.85
Valine	0.4	0.80
Cystine + methionine	0.5	0.72
Isoleucine	0.5	0.50
Leucine	0.6	1.20
Tyrosine + phenylalanine	0.5	1.30
Lysine	0.8	1.12
Histidine	0.2	0.35
Arginine	0.2	1.00
Tryptophan	0.2	0.20

ance. The requirements of the animals for individual amino acids are becoming better understood, although the divergence between sets of published figures is still too great for complete confidence. Differences may be due to the different techniques used for estimating requirements and it is known that other dietary constituents – particularly the levels of other amino acids, both essential and non-essential – may influence the requirement for a given amino acid (Harper, 1964). For example, the plant breeders might well be persuaded to attempt to increase the lysine content of an oilseed in order to render it more suitable as a non-ruminant feed. However, should the arginine content of the total diet in which the 'improved' oilseed is to be included not also be controlled, the effect of the additional lysine might be nullified. The values given in Table 19.2 are 'best estimates' based upon various sets of requirement figures published during the past decade. (For chicks see NRC, 1960, 1971; Dean & Scott, 1962, 1965; ARC, 1963; Dobson *et al.*, 1964; Payne & Lewis, 1966; Lewis, 1967; Fisher, 1972; Packham & Payne, 1972. For pigs see ARC, 1967; NRC, 1968.) Accumulating evidence strongly suggests that small changes in amino acid composition can have relatively large effects upon nutritive value for the non-ruminant.

Irregularities may be smoothed out up to a point by the addition of free amino acids to the diet, but this can be an expensive operation on a large-scale and may be inefficient if the absorption of synthetic amino acids differs from that of the same amino acid present as part of a normal protein chain. In consequence it seems likely that the best results are to be achieved by judicious combinations of natural protein feedstuffs, with additions of single amino acids kept to a minimum.

Lysine and methionine currently (March, 1973) cost £1.90 and £1.60 per kg respectively. Thus, their addition to a 'broiler starter' or a 'pig grower' diet costing respectively £75 and £60 per tonne, at a level equivalent to 0.1 % of the diet, would cost around £2 per tonne. Amino acid deficits will depend of course upon the formulation of the diet, but with an increasing reliance being placed upon plant protein sources owing to the rapidly escalating cost of animal protein, deficits of 0.2 % are possible. In the present world situation of protein shortage, the costs of protein supplements for animal feeding are at an unprecedented level. Groundnut and soya obtainable in 1971 for around £50 per tonne cost £125 in 1973, and fish meal has risen in the same period from £70 to around £200. It can easily be calculated that to provide 0.2 % of methionine using fish, soya or groundnut meal now would cost £13–17, and to provide 0.2 % lysine, £5–10 (Table 19.3).

The disparity between the price of synthetic amino acids and the price of protein concentrates may not be always maintained and there is, therefore, a good case for extending the range of available vegetable protein sources in order to provide the flexibility desired by the compounder of feedstuffs for pigs and poultry. Clearly a high-methionine soybean, a high-lysine sesame (*Sesamum indicum*) and cereals high in both of these amino acids are desirable providing that such strains can be developed without, at the same time, affecting other useful nutritional characteristics.

Some successes have already been achieved. High-lysine strains of maize (*Zea mays*) i.e., Opaque-2, and barley (*Hordeum vulgare*) i.e., Hiproly, have been evaluated in feeding experiments in recent years and found to be satisfactory. A study of a range of commercially available barleys has demonstrated significant differences in amino acid composition which are related to nutritive value (Woodham *et al.*, 1972*a*, *b*). For oilseeds Chopra has recorded that the lysine content of a range of groundnuts grown in India and the United States ranged from 2.47 to 4.20 g/16 g N (A. K. Chopra, personal communication). Differences in the amino acid composition of groundnuts grown in different localities have also been reported by Dawson & McIntosh (1973), and Wessels (1967) reported that chick diets containing Valencia groundnuts responded to threonine supplementation whilst diets containing other varieties did not. These differences are not due to differences in the technique of analysis. For rapeseed it has been shown that Polish and Argentine seed differs in its lysine content and in its response to heat treatment (Gray *et al.*, 1957). Deyoe & Shellenberger (1965) reported a

Table 19.3. *The cost of adding 0.2 % methionine or 0.2 % lysine to 1 tonne of feed, using fish, soya or groundnut meal (March 1973)*

	Cost/tonne March 1973 (£)	N × 6.25 (%)	Methionine + cystine (g/16 g N)	Lysine (g/16 g N)	Weight to provide 2.0 kg methionine (kg)	Weight to provide 2.0 kg lysine (kg)	Cost of providing 2.0 kg methionine† (£)	Cost of providing 2.0 kg lysine† (£)
Fishmeal	200	65	2.8	7.3	110	42	17.3	6.6
Soybean meal	125	45	2.7	6.8	164	65	13.7	5.5
Groundnut meal	125	50	2.0	3.2	200	125	16.7	10.4

† Cost corrected for cereal replaced at £40/tonne.

difference in the amino acid content of sorghum strains grown in different localities. The variations between many of the published analyses for all types of oilseed may be due in part to differences in analytical techniques, but it seems likely that at least some of the differences are real. We are not yet in a position to quote overall ranges of comparison for a given protein-containing food, but real differences exist and can be modified by plant breeding.

The problem of toxic seed components may also be solved by selection of appropriate strains. The discovery of glandless cottonseed free from gossypol seems likely to be of considerable importance for non-ruminant nutrition providing that an improved degree of disease-resistance can be bred into the strain. The good nutritional value of the material has already been demonstrated (Fisher & Quisenberry, 1971).

While much can be learnt from amino acid analysis regarding the suitability of a protein feedstuff for non-ruminants, such information must be reinforced by adequate toxicological studies in the case of novel materials. The problem of the cruciferous oilseeds has already been mentioned. Leaf protein concentrate on the other hand is a material whose potential seems to be correctly indicated by its amino acid composition. Yeasts, however, and particularly those cultured on hydrocarbon oil, have failed to fulfil the promise of an apparently good amino acid spectrum (Woodham & Deans, 1971). No toxic material has been detected and the explanation must be either the presence of a growth inhibiting substance or the non-availability of a proportion of the total protein. It has been suggested that the nucleic acids, which are not insignificant components of the yeasts, may have little or no nutritive value, and may even have a growth inhibiting effect. This awaits confirmation.

An attempt has been made in this compilation to assess the place of the non-ruminant animal both as a demanding and discriminating consumer of protein and also as a provider itself of high quality protein for feeding man. Much work remains to be done if pigs and fowls are to continue in their current role, and only if optimum efficiency of conversion of plant protein is achieved can the use of such raw material be countenanced. Even then it is to be feared that world conditions may eventually call for the gradual 'phasing-out' of non-ruminants as edible protein sources.

References

ARC (Agricultural Research Council) (1963). *Nutrient requirements of farm livestock*. No. 1: *Poultry*. HMSO, London.

ARC (Agricultural Research Council) (1967). *Nutrient requirements of farm livestock*. No. 3: *Pigs*. HMSO, London.

Boyd, D. A. (1951). *British Journal of Nutrition*, **5**, 255.

Byerly, T. C. (1967). *Science*, **157**, 890.

Dawson, R. & McIntosh, A. D. (1973). *Journal of the Science of Food and Agriculture*, **24**, 1217.

Dean, W. F. & Scott, H. M. (1962). *Poultry Science*, **41**, 1640.

Dean, W. F. & Scott, H. M. (1965). *Poultry Science*, **44**, 803.

Deyoe, C. W. & Shellenberger, J. A. (1965). *Journal of Agricultural and Food Chemistry*, **13**, 446.

Dobson, D. C., Anderson, J. O. & Warnick, R. E. (1964). *Journal of Nutrition*, **82**, 67.

FAO (1966–71). *Production yearbooks*, vols. 20–5. FAO, Rome.

Fisher, H. (1973). Methods of protein evaluation: assays with chicks and rabbits. In: *Proteins in human nutrition* (ed. J. W. G. Porter & B. A. Rolls), p. 263. Academic Press, London and New York.

Fisher, H. & Quisenberry, J. H. (1971). *Poultry Science*, **50**, 1197.

Frölich, A. (1953). *Kungliga Lantbrukshögskolans Anneler, Uppsala*, **20**, 105.

Gray, J. A., Hill, D. C. & Branion, H. D. (1957). *Poultry Science*, **36**, 1193.

Halnan, E. T. (1941). *Nature, London*, **148**, 336.

Harper, A. E. (1964). In: *Mammalian protein metabolism*, vol. 11 (ed. H. N. Munro & J. B. Allison), p. 87. Academic Press, London and New York.

Kirk, L. D., Mustakas, G. C. & Griffin, E. L. (1966). *Journal of the American Oil Chemists' Society*, **43**, 334, 550.

Leitch, I. & Godden, W. (1953). Imperial Bureau of Animal Nutrition Technical Communication No. 14, 2nd edition, p. 46. Aberdeen, Scotland.

Lewis, D. (1967). In: Proceedings of the nutrition conference for feed manufacturers, p. 24. Nottingham.

Lodge, G. A. (1970). Quantitative and qualitative control of proteins in meat animals. In: *Proteins as human food* (ed. R. A. Lawrie), p. 41. Butterworths, London.

Mann, G. E., Carter, F. L., Frampton, V. L., Watts, A. B. & Johnson, C. (1962). *Journal of the American Oil Chemists' Society*, **39**, 86.

Mustakas, G. C., Kirk, L. D., Griffin, E. L. & Clanton, D. C. (1968). *Journal of the American Oil Chemists' Society*, **45**, 53.

NRC (National Research Council) (1960, 1971). *Nutrient requirements of domestic animals*. No. 1: *Nutrient requirements of poultry*, 6th edition. National Academy of Sciences, Washington, DC.

NRC (National Research Council) (1968). *Nutrient requirements of domestic animals*. No. 2. *Nutrient requirements of swine*, 6th edition. National Academy of Sciences, Washington, DC.

Packham, R. G. & Payne, C. G. (1972). Proceedings of the 1972 Australian poultry science convention, p. 109.
Panel on the World Food Supply (1967). *The world food problem.* Report of the President's Science Advisory Committee, vol. 2, p. 338. The White House, Washington, DC.
Payne, C. G. & Lewis, D. (1966). *British Poultry Science*, **7**, 199.
Pons, W. A. & Eaves, P. H. (1967). *Journal of the American Oil Chemists' Society*, **44**, 460.
Wessels, J. P. H. (1967). *South African Journal of Agricultural Science*, **10**, 113.
Woodham, A. A. & Deans, P. S. (1971). *Proceedings of the Nutrition Society*, **30**, 59A.
Woodham, A. A., Savić, S., Ayyash, B. J. & Gordon, S. J. (1972*b*). *Journal of the Science of Food and Agriculture*, **23**, 1055.
Woodham, A. A., Savić, S. & Hepburn, W. R. (1972*a*). *Journal of the Science of Food and Agriculture*, **23**, 1045.

20. The conversion of animal products such as wool and feathers into food

F. B. SHORLAND

In the year ending June 1972 world production of clean wool, which is approximately pure protein, is estimated (New Zealand Meat and Wool Board, 1971–2) at 1.55 megatonnes (Mt) compared with 0.53 Mt of protein from lamb, mutton and goat meat assuming a protein content of 11 % (USDA, 1972). From the estimated production of poultry meat of 17.4 Mt in 1970 (FAO, 1971) it may be calculated (assuming the dressed carcass weight i.e., the meat plus edible viscera, and the feathers comprise 51 % and 8.4 % of the production respectively (Stewart & Abbott, 1961) and that their respective protein contents are 20 and 90 % (McCasland & Richardson, 1966)) that the meat and edible offal would yield 1.77 Mt of protein compared with 1.31 Mt available from the feathers. The relatively high contribution of wool as a protein source compared with lamb and mutton has previously been emphasised by Shorland (1968). Similarly Wilder (1953) pointed out in 1952 that the 600×10^6 broilers produced annually in USA would yield 64000 tons of feathers. In the conversion of feed protein to animal protein keratin production is invariably disregarded. Wilke (1966), for example, records 18 % efficiency for broilers in the conversion of protein ingested to edible protein. However, should the feathers be included the percentage efficiency would be raised to 32.

Wool offers the prospect of two-fold utilisation involving in the first place woollen clothing and carpet manufacture and in the second place recovery of the wool from these products for animal feeding. In the long term there is the possibility of direct conversion of both feathers and wool to human food.

The amino acid composition of keratins

It will be seen from Table 20.1 that the keratins listed in comparison with the FAO reference protein are all deficient in methionine. Whereas the lysine content of cattle hair is adequate, wool is marginal and feathers notably deficient. Tryptophan appears to be marginally deficient

Table 20.1. *Essential amino acid composition of proteins of actual or potential importance to food production (g amino acid/100 g amino acids)*

	Wool		Cattle hair[c]		Hog hair[d]		Feathers[e]						
	Raw[a]	Processed[b]	Raw	Processed	Raw	Processed	Raw	Processed	Wheat[f]	Maize[g]	Soybean[h]	Lamb[j]	FAO[k]
Arginine†	9.8	—	9.9	9.8	9.3	8.0	8.0	7.5	4.2	4.8	—	6.9	—
Histidine	1.2	0.9	2.1	0.8	1.1	1.1	0.6	0.8	2.2	—	—	—	—
Isoleucine	3.7	3.8	3.4	4.0	3.6	4.3	6.4	5.3	4.2	6.4	4.8	4.8	4.2
Leucine	8.9	8.4	8.1	9.5	7.9	8.0	8.5	9.2	7.0	15.0	7.3	7.4	4.8
Lysine	3.3	3.8	5.1	3.6	3.5	2.9	1.8	2.2	1.9	2.3	5.8	7.7	4.2
Methionine	0.6	0.6	0.4	0.4	0.7	0.7	0.5	0.5	1.5	3.1	1.4	2.3	2.2
Phenylalanine	4.0	3.5	1.9	2.1	2.8	3.4	5.5	5.6	5.5	5.0	4.8	3.9	2.8
Tyrosine	5.5	4.0	2.4	1.9	3.5	3.3	2.3	3.0	—	6.0	3.0	3.2	2.8
Threonine	6.5	6.4	7.1	7.2	6.0	5.2	4.6	4.4	2.7	3.7	3.8	4.9	2.8
Tryptophan	0.9	—	—	—	—	—	0.7	0.7	0.8	0.6	1.7	1.3	1.4
Valine	5.7	5.3	5.2	5.9	5.8	6.5	8.9	9.2	4.1	5.3	5.0	5.0	4.2
Cystine	5.2	5.9	5.3	2.9	11.0	3.5	8.7	3.3	—	4.8	—	1.3	2.0

† Marginal requirement.
References: [a] Corfield & Robson (1955); [b] Shorland & Gray (1970); [c] Moran & Summers (1968); [d] Moran, Bayley & Summers (1967); [e] Block & Bolling (1951); [f] Masek (1966); [g] Wilke (1966); [h] Kohler (1966); [j] American Meat Research Foundation (1964); [k] FAO (1957).

in all keratins. Some of the vegetable sources are likewise shown to be deficient in essential amino acids. Wheat and maize are low in lysine and wheat in addition lacks methionine. Whereas chemical processing to render wool digestible has little effect on the amino acid composition, heat processing of feathers destroys some cystine. In addition to the amino acids listed in the FAO reference protein, histidine, which is required by growing rats at the level of 2.3 g/100 g amino acids (NRC, 1963), is deficient in most keratins. However, the requirements for glycine by young turkeys and chickens (Long, 1961) would generally be met from these sources.

The conversion of keratins to digestible proteins

Treatments for rendering keratins digestible include conversion to finely divided powder by grinding, reaction with chemical agents, use of enzymes and heat treatment. The commercial utilisation of keratins up to the present time has been mainly confined to hydrolysed (autoclaved) feather meal which is used as a partial replacement for meat meal and other proteins in practical rations for poultry feeding (cf. Ewing, 1963).

Effect of fineness of grinding

Routh & Lewis (1938) found that wool after grinding to a powder in a ball mill was readily digested by trypsin and pepsin. This result was in agreement with Kuhne's observation made some 60 years earlier (Kuhne, 1878), that hair became digestible with pepsin when the surface area was increased by mechanical means. It was similarly found by Olcott (1943) that hoofs from cattle, horses and pigs after grinding to a powder were attacked by pancreatin. Routh (1940) noted that in the preparation of powdered wool in a ball mill the cystine content was lowered and inorganic sulphate produced. Newell & Elvehjem (1947) found that a process which pulverised the keratin to a fine powder without heating produced a product of higher nutritive value than a powder of comparable fineness made by ball milling in which heat was generated. It would appear therefore that the increased digestibility of keratins is dependent on the fineness of grinding *per se* and not on the chemical changes induced by grinding.

It has been established by Routh (1941) that powdered wool as the sole source of protein in an otherwise adequate diet at the level of 15 % and after fortification with tryptophan, methionine, histidine and lysine supported moderate growth (1.5 g/day) in weanling rats. Similar results were obtained with powdered chicken feathers (Routh, 1942). In diets supplemented with 5–8 % casein no differences in nutritive value between powdered wool or feathers were found but with 3 % casein, powdered wool produced a higher growth rate.

According to Wagner & Elvehjem (1942) chicks and rats fed finely ground swine hoofs as a source of protein, at the level of 24 % and 30 % of the diet respectively, grew normally. Similarly they found (Wagner & Elvehjem, 1943) that powdered swine hoofs were satisfactory as a substitute for meat meal and fish scraps in practical starting rations for chicks. It was further shown (Newell & Elvehjem, 1947) that chicks and rats fed finely ground hog hair at high levels as a sole source of protein exhibited moderate growth rates whereas finely ground chicken feathers allowed only poor growth.

In contrast to the above results Slinger *et al.* (1944) reported that finely ground (75 % passing a 100 mesh sieve) mixed hoof and horn keratin when fed to chicks had no value in combination with commonly used vegetable proteins sources such as soybean meal.

Chemical processing

In the patent literature, which is not reviewed in detail here, references may be found to procedures for the extraction of protein from keratin by chemical methods. Anker (1969), for example, has recently proposed the use of sodium sulphide (Na_2S) to recover protein from keratin followed by precipitation at pH 4.2 with 6 N HCl. Similar procedures have previously been used by Koerner *et al.* (1952). Other procedures include heating keratins such as feathers in an aqueous medium buffered to pH 8–10 in a closed system at 115–35 °C for 10–30 min (Weeks, 1971); and digestion in an aqueous medium containing 0.1–5.0 % keratinase and 0.5–3.0 % reducing agents such as $HSCH_2.CH_2OH$ and Na_2SO_3 at 10–70 °C (Weeks & Wildi, 1970).

The effectiveness of treatment with sodium sulphide to render keratins digestible was demonstrated by Draper (1944). He showed that Na_2S-treated feathers as well as autoclaved and ground feathers, when added to a basal cereal diet, resulted in a growth rate in chicks and rats greater than that produced by the basal cereal diet.

Moran *et al.* (1966) treated feathers with such reducing agents as sodium thioglycollate and sodium sulphide using 1/4 the molar cystine equivalent on the basis that four moles of cystine per 10400 g of feather keratin were present. The reaction mixture was raised to pH 11 and after standing for 3 h lowered to pH 6 by adding hydrochloric acid. The precipitated meal was filtered and washed with distilled water and dried. It was found that the meal prepared with the lowest concentrations of the reducing agent when supplemented with methionine, histidine, tryptophan, lysine and glycine and fed at the 15 % level as the sole source of protein gave a growth response similar to that of soybean meal or autoclaved feather meal similarly fortified. When higher levels of sodium thioglycollate were used in the preparation of the feather meal, the growth response was depressed suggesting the presence of toxic factors. In another experiment, which included a preliminary treatment of the feathers with 0.1 N NaOH at pH 11 for 12 h, the preparation failed to produce any growth response. This suggests that the nutritive value of the product is affected by the conditions of its preparation.

Shorland & Matthews (1968) and Shorland & Gray (1969, 1970), by paying particular attention to the optimum conditions in preparing digestible wool protein (SWP) by the method outlined by Koerner *et al.* (1952), were able to increase the yield to 91 % of the amount present in the untreated wool. SWP was also prepared by means of papain (Shorland & Bentley, 1969) following the procedures outlined by Crewther *et al.* (1965) having regard to the optimum conditions of the reaction as outlined by Lennox (1952).

The nutritive value of SWP was tested on weanling rats of the Wistar strain at the level of 10 % as the sole source of protein in a diet that was otherwise adequate. Without amino acid supplementation, SWP failed to support growth. As shown in Table 20.1 apart from the methionine, which varied from 0.49 to 0.66 g/100 g protein, SWP generally met the requirements of the FAO reference protein but the levels of isoleucine and lysine were marginal. To meet the special needs of weanling rats, histidine was added to raise the level to 2.3 g/100 g protein (NRC, 1963). A growth gain of 1.6 g/day was achieved by the addition of methionine alone to the level of 2.0 g/100 g amino acids. The addition of tryptophan was without effect, but when lysine was added to bring the total content to 6.0 g/100 g the gain increased to 2.0 g/day providing a PER value of 1.8. This may be compared with values of 4.0 for dried defatted egg powder, 2.28 for dried skim milk and −0.10 for bread (Bender, 1956). Similar results were obtained for solubilised off-cuts from dyed woollen cloth and for the SWP prepared by means of papain. It would appear that as compared with the powdered keratins described above SWP is more nearly complete with respect to its essential amino acid content.

The recent work of Moran and co-workers (1966) has established that hydrolysed feather meal (autoclaved at 142 °C for 30 min) fortified with DL-methionine, L-lysine, L-histidine, L-tryptophan and glycine when used as the sole source of protein at the level of 15 % in an otherwise adequate diet for chicks gives growth gains equal to that of soybean protein. Similar results were obtained with hydrolysed hog hair (Moran *et al.*, 1967) and cattle hair (Moran & Summers, 1968). McCasland & Richardson (1966) found that in line with the results obtained for chicks, hydrolysed feather meal fed to weanling rats at the level of 10 % and 20 % respectively when supplemented with lysine, methionine, tryptophan and histidine supported a rate of growth slightly less than the equivalent amount of purified soybean protein. They also showed that the nutritive value of hydrolysed feather meal as measured by growth of rats, pepsin digestion, and quantitative microscopy of faeces of rats, agreed closely.

Utilisation of keratins as human food

Shorland & Bentley (1968, 1969) examined the feasibility of replacing ordinary flour with wool flour up to the extent of two-thirds. For these tests wool flour prepared by solubilisation with papain was generally used because solubilisation by Na_2S–$NaHSO_3$ required careful removal of residual sulphur and other by-products to ensure absence of undesirable flavours. To ensure acceptability it was necessary to grind the wool flour to the same degree of fineness as ordinary flour to eliminate grittiness.

Conventional foods such as sponge cakes, scones, bread (wholemeal) and ginger nuts were tested in which ordinary flour was partially replaced not only by wool flour but also other protein flours such as soybean flour, casein and flour prepared from grass and tripe.

Sponge cakes in which the above-mentioned protein flours replaced conventional flour to the extent of 37 % (by weight) were as acceptable as conventional sponges. The appearance of residual chlorophyll in the 'grass' sponge, despite extraction of the flour with 95 % alcohol, lowered acceptability when the tests were made without prior blindfolding. Further detailed testing of the 'wool' sponge showed that in a panel of twelve tasters comparing the 'wool' sponge with the conventional, ten tasters found no difference and two actually preferred the wool sponge. Similar results were obtained with scones but in one batch of wool protein prepared chemically, a sulphurous taste was noted which resulted from incomplete washing to remove contaminants. The texture of 'wool' bread appeared a little firmer which to some made it less acceptable than the conventional product but to others more acceptable.

Tests on the upper limit of substitution of wheat flour by wool protein flour showed that in ginger biscuits at the 60 % level of replacement by SWP flour, the product became less acceptable. With complete substitution of wheat flour by wool flour there was a loss of cohesion in the ginger biscuit, suggesting that wheat flour contributes more to the binding of ingredients than to the flavour of the product.

From the above investigations, guidelines for the manufacture of edible nutritous protein from keratin have been set. There seems no doubt that with the addition of methionine to solubilised wool protein and of this amino acid together with lysine and tryptophan to solubilised feather protein, at the cost of a few cents per kilogram, the way is clear for their eventual use directly in human diets.

References

AMRF (1964). Summary of nutrient content of meat. American Meat Research Foundation Bulletin No. 57.

Anker, C. A. (1969). British Patent No. 1 143 556.

Bender, A. E. (1969). *British Journal of Nutrition*, **10**, 135–43.

Block, R. J. & Bolling, D. (1951). *The amino acid composition of proteins and foods*. Charles C. Thomas, Springfield, Illinois.

Corfield, M. C. & Robson, A. (1955). *Biochemical Journal*, **59**, 62–8.

Crewther, W. G., Fraser, R. D. B., Lennox, F. G. & Lindley, H. (1965). The chemistry of keratins. *Advances in Protein Chemistry*, **20**, 191–346.

Draper, C. I. (1944). Iowa Agricultural Experiment Station Research Bulletin No. 326.

Ewing, W. R. (1963). *Poultry nutrition*, 5th edition. Ray Ewing Co., Pasadena, California.

FAO (1957). FAO nutrition studies No. 16. FAO, Rome.

FAO (1971). *Production yearbook*. FAO, Rome.

Koerner, E. C., Ehrhardt, H., Haigh, P. & Kirchof, J. (1962). US Patent No. 2591945.

Kohler, G. O. (1966). Safflower; a potential source of human food. In: *World protein resources*. Advances in Chemistry series No. 57. American Chemical Society, Washington, DC.

Kuhne, W. (1878). *Untersuchungen der Heidelberg Physiologischen Institut*, **1**, 219. (Quoted by Routh, 1940.)

Lennox, F. G. (1952). *Australian Journal of Scientific and Industrial Research*, **5**, 189–209.

Long, C. (1961). *Biochemists' handbook*. E. & F. Spon, London.

McCasland, W. E. & Richardson, L. R. (1966). *Poultry Science*, **45**, 1231–6.

Masek, J. (1966). *Proceedings of the International Congress on Nutrition*, **4**, 780–96.

Moran, E. T., Jr, Bayley, H. S. & Summers, J. D. (1967). *Poultry Science*, **46**, 548–53.
Moran, E. T., Jr & Summers, J. D. (1968). *Poultry Science*, **47**, 570–6.
Moran, E. T., Jr, Summers, J. D. & Slinger, S. J. (1966). *Poultry Science*, **45**, 1257–66.
Moran, E. T., Jr, Summers, J. D. & Slinger, S. J. (1967). *Poultry Science*, **46**, 456–65.
NRC (1963). National Research Council Publications No. 1100. Washington, DC.
Newell, G. W. & Elvehjem, C. A. (1947). *Journal of Nutrition*, **33**, 673–83.
New Zealand Meat & Wool Board Economic Service (1971–2). Annual Review of the sheep industry. Publication No. 1595.
Olcott, H. S. (1943). *Proceedings of the Society for Experimental Biology and Medicine*, **54**, 210.
Routh, J. I. (1940). *Journal of Biological Chemistry*, **135**, 175–81.
Routh, J. I. (1941). *Journal of Nutrition*, **23**, 125–30.
Routh, J. I. (1942). *Journal of Nutrition*, **24**, 399–404.
Routh, J. I. & Lewis, H. B. (1938). *Journal of Biological Chemistry*, **124**, 725–31.
Shorland, F. B. (1968). *Food Technology in New Zealand*, **3**, 10–11.
Shorland, F. B. & Bentley, K. W. (1968). *New Zealand Journal of Science*, **11**, 722.
Shorland, F. B. & Bentley, K. W. (1969). *Food Technology, Australia*, **21**, 218–20.
Shorland, F. B. & Gray, J. M. (1969). *New Zealand Journal of Science*, **12**, 647.
Shorland, F. B. & Gray, J. M. (1970). *British Journal of Nutrition*, **24**, 717–25.
Shorland, J. B. & Matthews, J. R. (1968). *New Zealand Journal of Science*, **11**, 131–6.
Slinger, S. J., Evans, E. V., Kellam, W. I. & Marcellus, F. N. (1944). *Poultry Science*, **23**, 431–6.
Stewart, G. F. & Abbott, J. C. (1961). Marketing guide No. 4. FAO, Rome.
USDA (1972). Foreign agriculture circular: livestock and meat consumption. United States Department of Agriculture, Washington, DC.
Wagner, J. R. & Elvehjem, C. A. (1942). *Proceedings of the Society for Experimental Biology and Medicine*, **51**, 394–6.
Wagner, J. R. & Elvehjem, C. A. (1943). *Poultry Science*, **22**, 275–7.
Weeks, L. E. (1971). French Patent No. 2065014.
Weeks, L. E. & Wildi, B. S. (1970). German Patent No. 1962207.
Wilder, O. H. M. (1953). *Poultry processing and Marketing*, **59**, 11, 27.
Wilke, H. L. (1966). General outlook for animal protein in food supplies for developing areas. In: *World protein resources*. Advances in Chemistry series No. 57. American Chemical Society, Washington, DC.

21. Increasing the direct consumption of fish

G. H. O. BURGESS

Fish as a food

Nobody who goes to sea on a trawler and sees the contents of the net spilled on to the deck can fail to be impressed by the enormous variety of marine life. The term fish is often loosely used to include not only the teleost and elasmobranch fishes, two vertebrate classes with very different structure and composition, but also shellfish, generally understood to be the Crustacea and Mollusca. Furthermore, there are many relatively local fisheries for other forms of marine life from echinoderms (sea urchins, bêche-de-mer) to lampreys, representatives of the primitive vertebrate class of cyclostomes, and even polychaetes (Palolo worms).

Clearly it is impossible to deal in detail with such variety in a single chapter and attention will be given here mainly to the teleost (bony) fish, which in 1971 made up over 90 % of the world catch; the elasmobranches such as sharks, rays and chimaeras, comprised well under 1 %, and the rest were crustaceans and molluscs (7 %), aquatic mammals and plants. Even with this restriction, the field is a formidable one. In the area of north-west Europe alone roughly 400 species of fish, mostly teleosts, are found in lakes, rivers and seas, and about 60 species are caught commercially for food. Throughout the world hundreds or even thousands of species are eaten and it is difficult to generalise about their composition and nutritional properties.

The nutritive value of fish proteins appears to be at least equal, and perhaps slightly superior, to that of land animals (Connell, 1970). The composition of fish muscle varies, however, both from species to species and within species from one season to another (Burgess & Shewan, 1970).

Figures such as those given in Table 21.1, however, have little value beyond illustrating that proximate composition can vary greatly and figures for average composition of fish flesh are almost meaningless. Within one species the amount of protein in the flesh is generally fairly constant, though severe stress, such as spawning after a period of prolonged starvation, can result in a high water content; in extreme cases this can be over 90 %.

Table 21.1. *Proximate composition of edible portion of fish* (from Stansby & Olcott, 1963)

Statistic calculated	Moisture %	Protein %	Oil %	Ash %
Average	74.8	19	5	1.2
Range	28–90	6–28	0.2–64	0.4–1.5
Ratio high to low	3.2	4.7	320	3.8

The lipid content of fish flesh varies according to species. Some, such as cod (*Gadus morrhua*), contain little fat in the flesh, the reserves being mainly in the liver. Other species, such as herring (*Clupea harengus*) may contain 25 % fat or more in the flesh. Lovern & Wood (1937) demonstrated in a classic paper, however, that the lipid content changed dramatically with season and similar observations have been made in other species. The depot fats of most teleosts are triglycerides, though a few species have wax esters. Elasmobranchs, as in other respects, are peculiar in their lipid metabolism and biochemistry; some utilise alkoxydiglycerides, generally with triglycerides. The hydrocarbon squalene also occurs in considerable quantities in the liver oil of some species (Lovern, 1962). Teleost depot lipids typically contain complex mixtures of fatty acids, ranging in chain length from 14 to 22 carbon atoms, and which include many highly-unsaturated compounds. The ease with which these fats are oxidised means that certain measures must be taken in the handling and processing of fatty species for food if they are not to become unacceptably rancid.

Fish may contain Vitamins A and D and a number of water soluble vitamins of the B group. Vitamins A and D are generally to be found, where they occur, in the visceral fat, though a few species contain appreciable amounts in the flesh. The lamprey appears quite exceptional however; according to Higashi (1962) one species contains from 14000 to 63000 iu Vitamin A (4.2 to 18.9 mg) per 100 g flesh. Liver oils have the largest potency of Vitamin D, some species of tuna containing from 25000 to 250000 iu (0.625 to 6.25 mg Vitamin D_3 equivalent) per gram of oil (Higashi, 1961). Little Vitamin D occurs in the flesh. Some fish are useful sources of vitamins (Table 21.2) but they may be greatly affected by processing.

The dark muscle is most marked in those species which swim for long periods at a time. Species such as cod which swim actively for comparatively short periods at a time, possess relatively small amounts of

Table 21.2. *Content of water soluble vitamins in fish flesh* (from Higashi, 1962)

Species	Part of flesh	B_1 (μg/g)	B_2 (μg/g)	B_6 (μg/g)	B_{12} (μg/100 g)	PA (μg/g)	FA (μg/100 g)	Niacin (mg/100 g)
Mackerel	White	0.8	0.3	14	0.9	1.6	1.9	21.4
	Dark	8.8	2.8	10	8.0	16	5.8	19.8
Horse mackerel	White	1.5	0.8	—	0.3	—	—	—
	Dark	3.3	7.8	—	7.8	—	—	—
Skipjack	White	0.6	0.3	10.5	2.5	—	—	24.5
	Dark	5.2	6.9	9.0	16.5	—	—	12.2

PA, pantothenic acid; FA, folic acid.

dark muscle. Braekkan (1962) gives a full summary of work on B vitamins in fish and shellfish.

The shell of molluscs and crustaceans can constitute a considerable proportion of the total weight. Crustacean protein compares well with casein, beef and egg albumen. The total protein content is usually lower in molluscan tissue; perhaps about half that in crustacea. Rather more than muscle tissue alone is eaten when shellfish are consumed, however, and some discrepancies in figures arise because analyses are not all carried out on the same basis (see Borgstrom, 1962, for a full discussion of this topic).

The resource and its present utilisation

The world catch of marine fish has been steadily increasing over the last two decades and in 1971, the latest year for which figures are available, was over 69 Mt (Fig. 21.1). Estimates of the maximum sustainable yield vary considerably; not only are there many uncertainties inherent in making predictions of this kind but there is also the additional complication of whether to include resources in the estimate that are at present scarcely utilised, or not at all, yet might in future be turned to account in some way. An outstanding example of such a resource, which owing to difficulties of harvesting and processing is at present virtually unused, is Antarctic krill. An estimate of the potential annual yield of one species of krill alone, *Euphausia superba*, is in excess of 50 megatonnes (Gulland, 1971). According to the same source, the potential yield of what may be termed the traditional fishes is estimated to be about 100 megatonnes and it is interesting that ten years ago a forecast was made of 115 Mt as the potential total harvest of

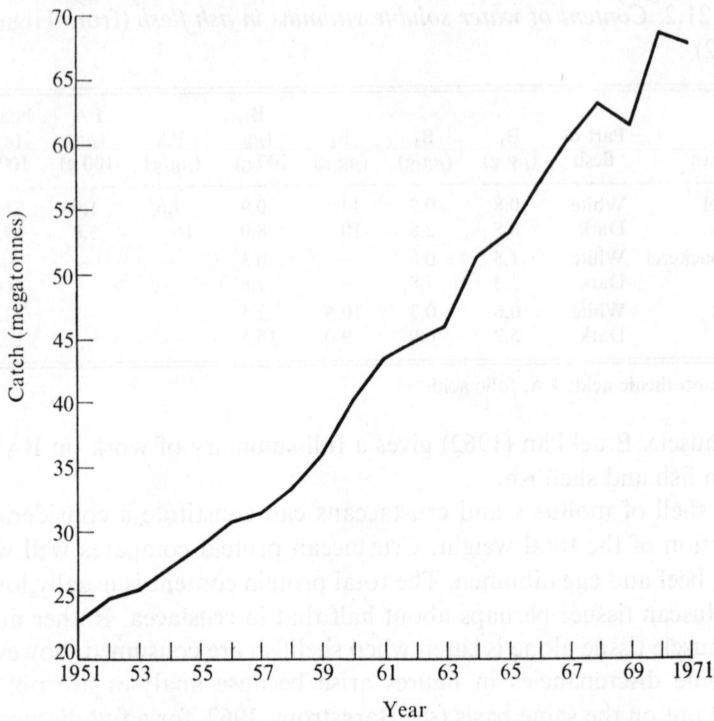

Fig. 21.1. World nominal catch 1951–71 (FAO, 1964, 1972).

marine fish each year (Graham & Edwards, 1962). It may be expected that the rate of increase in the world catch will fall during the next few years unless ways are found of utilising non-traditional resources, and indeed there seem to be indications that this trend has already begun. In 1971 about 64 % of the world landing was caught for direct human consumption whilst 36 % was utilised for making fish meal and similar products, mainly for animal feeding. A significant proportion of the catch for human consumption was also converted to meal since not all the parts of a fish as landed are regarded as edible. Roughly half the weight of a gutted cod (*Gadus morrhua*), for example, is accounted for by head and backbone; these are removed by filleting. This so-called offal made up 659 kilotonnes (kt) of the 1971 Japanese catch of 9908 kt (6.7 %) and, also in 1971, offal in the United Kingdom amounted to 368 kt in a total catch of 1089 kt (33.8 %) (FAO, 1972). Such differences in utilisation from country to country arise not only from dissimilarities in the species landed but also in national eating habits and, in some instances, the efficiency of utilisation of the offal.

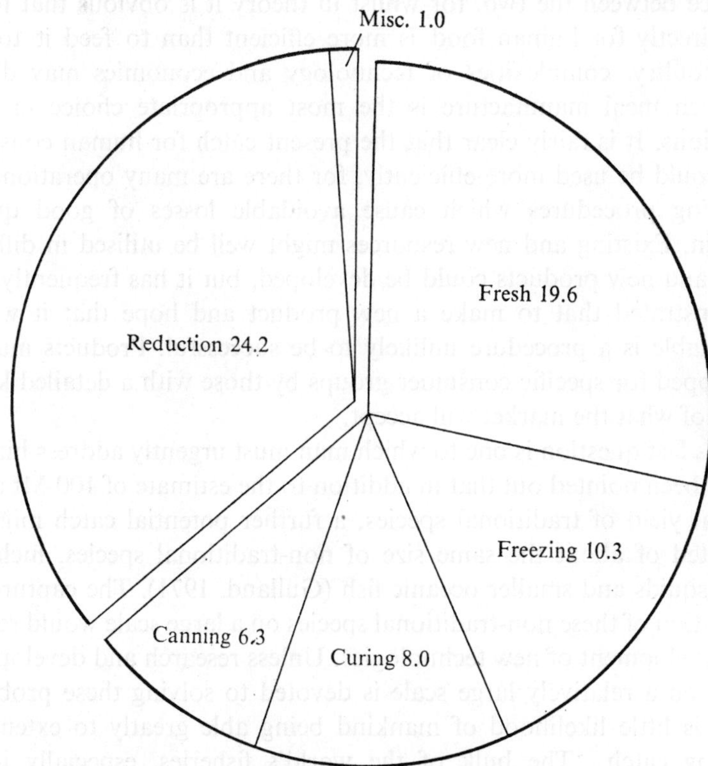

Fig. 21.2. Disposition of 1971 world catch in megatonnes (FAO, 1972).

Fig. 21.2 shows the disposition of the 1971 world catch by types of product. The quantity of fish cured, that is preserved by traditional methods of drying, salting or smoking, has remained remarkably constant at around 8 Mt for the last decade. The amount canned, frozen and marketed fresh has been slowly rising and this trend seems likely to continue, probably, if experience in developed countries is any guide, with steadily more emphasis on the use of freezing and cold storage.

The present pattern of utilisation raises a number of questions about the most efficient use of the resource for human food. First there is the most appropriate level of exploitation and how catching rates can be controlled. These are not matters of direct relevance to this chapter but they are vital in determining how the resource shall be managed. The catch for human consumption has been rising but the industrial catch has been rising even faster; 33 % of the total world catch went for reduction in 1966 and over 36 % in 1971. It is not easy, however, to decide the balance between the two, for whilst in theory it is obvious that to use fish directly for human food is more efficient than to feed it to pigs and poultry, complexities of technology and economics may dictate that fish meal manufacture is the most appropriate choice in some situations. It is fairly clear that the present catch for human consumption could be used more efficiently, for there are many operations and handling procedures which cause avoidable losses of good quality protein. Existing and new resources might well be utilised in different ways and new products could be developed, but it has frequently been demonstrated that to make a new product and hope that it will be acceptable is a procedure unlikely to be successful. Products must be developed for specific consumer groups by those with a detailed knowledge of what the market will accept.

This last question is one to which man must urgently address himself. It has been pointed out that in addition to the estimate of 100 Mt as the annual yield of traditional species, a further potential catch might be expected of about the same size of non-traditional species, including krill, squids and smaller oceanic fish (Gulland, 1971). The capture and utilisation of these non-traditional species on a large scale would require the development of new technologies. Unless research and development effort on a relatively large scale is devoted to solving these problems, there is little likelihood of mankind being able greatly to extend the existing catch. 'The bulk of the world's fisheries, especially in developing countries, will remain based on the familiar types of fish. For these the pattern of expansion during the post-war years – a doubling every ten years – cannot continue for much longer.' (Gulland, 1971.)

Aquaculture

Mention should also be made of the possibilities of increasing the production and hence the utilisation of fish by aquaculture or, to use the more popular term, farming. It has been suggested that by 1985 there may be a fivefold increase in the amount of farmed fish (FAO, 1970) and perhaps as much as 30 Mt by the year 2000. Space will allow here only a brief comment on some of the more important factors, and the reader should consult one of the many works on the subject of fish culture for further information (e.g., Bardach *et al.*, 1972). Vertebrate freshwater and marine fish, molluscs and crustaceans, and other aquatic organisms, can be and indeed are cultured commercially in various parts of the world.

Lamellibranch molluscs such as oysters and mussels are filter feeders.

Their growth rates can be high, and they may reach marketable size in a year, but such rates depend upon an adequate supply of water rich in appropriate species of phytoplankton and at or near the optimum growing temperature for the species in question.

Many vertebrate species of fish are carnivorous and, for rapid growth, their diets must contain a considerable proportion of animal protein. Commercial diets usually contain at least 30 % on a dry weight basis and, for some species, up to twice this level. Fish meal is frequently used as a protein source because of its superior nutritional properties. In nature, of course, fish near the top of the food chain consume living organisms composed largely of protein, fat and water; modern feed formulations contain carbohydrate as an energy source, though the maintenance of a proper protein/energy balance is important for maximum growth rates (Lee & Putnam, 1973). In temperate areas, particularly in the western world, vertebrate fish farming is largely concerned with carnivorous species such as the salmonids, and particularly rainbow trout (*Salmo irideus*), and is therefore mainly a protein conversion business, in a sense in competition with the farming of pigs and poultry. It is not creating new protein. Efficiency of conversion, assuming a properly balanced diet, under optimum conditions can reach 3 or 4:1, all on a dry weight basis. The literature can be confusing, however, since some authors give the dry weight of feed required to give unit weight of live fish, on which basis figures of just over 1:1 are possible.

In tropical and sub-tropical areas and in Asia generally, it is the practice to rear freshwater herbivorous species such as *Tilapia*. Species of carp, which are omnivorous, have been grown in China for many centuries. In the Far East fish culture and agriculture are often carried out together, a crop of fish being produced in the flooded rice fields. More intensive techniques are also used. The high water temperatures and light intensities make it possible to obtain one or more crops a year of marketable fish. Yields depend on many factors, such as whether fertiliser or additional feed is employed, as well as the species involved. Hickling (1962) gives an example of ponds at Malacca which yielded a standing crop of only about 22 kg/ha·yr, with 55 kg as a maximum, yet after two years of treatment with fertiliser alone, crops of the order of 1680 kg/ha were obtained. The exceptionally high figure of 7530 kg/ha in a pond of 0.6 ha cultivated according to traditional Chinese methods is quoted by Hickling from Liu (1955).

Conversion ratios for herbivorous species are more difficult to assess. Bardach *et al.* (1972) give a conversion ratio, presumably on a wet weight basis, of 48:1 for *Tilapia* spp. in Zambia fed on a mixture of plant foods such as grass and chopped leaves of various plants such as banana, sweet potato and kale, but much more favourable conversion levels would no doubt be obtained on more intensive feeds.

Methods of utilisation

Management of fisheries, as this term is commonly used, often requires limitation of fishing effort, probably selectively; the objective of such management, at least in an ideal situation, is to control a fishery so that it approaches its maximum sustainable yield. This chapter will not discuss this aspect of management but is concerned rather with the question of what use can be made of fish once it is caught.

Spoilage of the catch, which begins as soon as fish die, is largely due to bacterial activity; associated catabolic changes due to enzymic degradation also occur and both are temperature dependent. Many general works describe these changes in more detail (Shewan 1961; Liston *et al.*, 1963; Burgess *et al.*, 1965). The precise nature of the bacterial flora depends upon the area in which fish is caught, the predominantly psychrophilic flora in Arctic and temperate waters giving way to mainly mesophilic types in sub-tropical and tropical waters. Traditional methods of preservation were mainly developed in prehistoric times and rely on basically simple techniques of salting, drying, fermentation or smoking. Modern methods are largely dependent on technological developments that have occurred within the last two centuries; canning requires the manufacture of containers which can be hermetically sealed and heat sterilised, and freezing or fresh fish marketing involve the use of mechanical refrigeration. Various new techniques such as freeze-drying or irradiation have been proposed but for a number of reasons have not yet been applied on any wide commercial scale. Some other developments, such as enzymic digestion of fish or the production of fish protein concentrate may have particular value in special situations. With such a wealth of choice of methods available the question arises of what is the best way of dealing with the catch from a particular area. On the other hand it may perhaps be better to turn it into fish meal. The answers to these problems cannot be given simply; there is no 'best' way of utilising this unique resource, and any decision must be reached after weighing the advantages and disadvantages of the various options available. A recent review deals with this more fully than is possible here (Burgess, 1971).

Traditional methods

Over 10 % of the world catch is still preserved by traditional methods such as salting and drying. Products usually keep well without requiring special methods of storage; manufacture, although often complicated and requiring special skills, involves only simple equipment; methods are usually well understood locally in the fisheries which depend on them. This implies that the product is adapted to local species of fish and, what is equally important, is locally acceptable.

Unfortunately, traditional products are often wasteful, either because the methods inevitably involve loss of nutrients or because in the primitive conditions of manufacture it is difficult to control losses due, for example, to insect infestation. A 30 % weight loss may occur in pickled salt cod (Beatty & Fougère, 1957) and although this is mainly water it also contains a few percent protein and presumably water-soluble vitamins and flavour constituents. Cooking losses, for example during hot smoking, may be 20 % or more by weight; fluid from pressure-cooked tuna contains 2.5 to 5 % solids (Dollar *et al.*, 1967) and losses from ordinary cooking operations are probably similar. Furthermore, the preparation of some traditional products involves dressing the fish; salt cod is made from split fish from which entrails, head and part of the backbone have been removed. Salt herring is made from partly gutted fish. Many products, however, are prepared from whole ungutted fish and some of the fermented products of SE Asia depend on the presence of gut enzymes for their successful preparation.

The preservation processes themselves can result in some nutritional losses, but it must be accepted that this may be a price well worth paying provided the fish reaches those in need. Nevertheless, traditional products are suitable mainly for healthy adults; they are usually unsuitable for weaned infants, the sick and the aged or, in other words, the groups most often at risk in impoverished communities.

Insect and rodent infestation, mould growth and bacterial attack can all contribute to the enormous losses which are inherent in the methods; in many instances these are unavoidable bearing in mind the high temperatures and humidities, lack of suitable wrapping materials and the inadequate storage facilities in particular areas.

Modern methods

These are mainly the landing and distribution of fresh fish, freezing and canning. Perhaps also to be included are special products such as fish sausage, acid pickled fish and fish pastes which will remain edible for a few days, weeks or even months provided they are adequately chilled. Fresh fish, of course, has always been available to those people on sea coasts, lakes and rivers but the distribution of fresh fish to other areas is a modern development.

The important points to notice are that really fresh and properly frozen fish are of high nutritional value. Fluid, containing about 5 % protein, may be lost from the tissue during storage of wet fish; losses in excess of 10 % have been quoted by various authors. Badly frozen and stored fish may lose 15 % or even more as 'drip' on thawing. These, of course, are extreme figures but since the liquid may contain 7 % protein, the losses are not negligible.

Although freezing and canning do not impair the nutritive value, these methods often use only the fleshiest parts; if fish is filleted as much as 50 % of the gutted weight may be wasted. The remainder may be used for fish meal and is not entirely lost, but nevertheless it could be argued that the best use is not being made of a first-class protein resource.

Modern products have wide acceptability and are suitable for all groups in the community. Unfortunately, however, their production and marketing require chilling, icemaking, freezing, cold storage and canning facilities; this implies not only skilled operators and maintenance engineers but also considerable capital investment.

The proper use of ice in tropical and sub-tropical areas can extend the range of fishing vessels and enable good quality fish to be landed, but freezing costs are higher than in temperate areas, more ice is required for the same weight of fish and losses through melting are higher even when vessels are well insulated. In some inshore fisheries vessels are too small to allow the use of ice. Furthermore, development of a fresh or frozen fish industry also requires a rapid distribution network. Fresh fish is perishable and frozen fish must be carried quickly from one cold store to another or to the consumer, in well insulated and refrigerated transport. Nevertheless, modern fish industries are being built up in a number of developing countries, for example West Africa.

It is, in fact, lack of capital investment rather than technological problems that prevent the wide development in some areas of modern fish industries. There seems no doubt that given the choice people prefer modern products to traditional ones. Some relatively minor technological problems remain to be solved but these are not insuperable.

Special methods

These are a miscellany of still largely experimental methods and include some, such as radiation pasteurisation, which are unlikely to find application for some time to come. Freeze-drying, also, although it perhaps gives a better quality product than that from conventional drying methods, is not so obviously superior as to commend itself. Neither causes serious change in the nutritional value of fish, though they affect palatability and it is on grounds of loss and safety (for irradiation), and cost (for freeze-drying), that these have not been developed commercially for fish on any scale.

The most promising lines of work have been in the development of various types of fish protein concentrate (FPC). In a recent review on FPCs, these have been defined as 'products obtained from fish bearing higher protein contents than the original raw material' (Knobl, 1973). FPCs range, therefore, from various types of fish meal, through products prepared by solvent extraction of fish, to enzymic digests. They have certain features in common, such as a lack of functional properties, and are regarded as protein supplements rather than as food in their own right.

In some parts of Africa, fish meal is acceptable as human food; it has a strong flavour which is relished in some areas. There have been various attempts to manufacture high quality and bacteriologically safe meals for this market, which could perhaps be expanded. Odourless, tasteless and bland FPC prepared by solvent extraction of fish, involving the use of isopropyl alcohol at some stage, has been studied for a decade or so, especially in North America. The technological problems of manufacture have been overcome but attempts to develop processing plants have mostly been thwarted, for many reasons including political and economic ones. FPCs have one great potential advantage over most other products; they can be made from small fish, such as Peruvian anchoveta, which would otherwise be unusable. Their limitations are mainly that they are supplements and therefore likely to find their greatest use in subsidised feeding programmes.

Enzymic digests of fish flesh can be used for the manufacture of a range of products. Hitherto they have been made commercially mainly as milk replacements in calf feeding, but they have potential elsewhere. Bones and oil are removed from the digest and after the addition of hardened fat, it is dried. The best products are white powders with slightly fishy flavour and good dispersion in water.

Why not fish meal?

In the long term it is clear that the manufacture of fish meal is not the best way of using fish that could be eaten directly by man; nevertheless, industrial fisheries have some virtues for developing countries. The establishment of an industrial fishery may, therefore, be regarded as one step in the setting up of a food fishery. Although the technology of meal production is fairly complex, plants are compact, they are sited at fishing ports and they require only a small number of trained people for their operation. Furthermore, there is a world famine of high grade protein for animal feed and thus a ready sale of the product in developed countries. It is, therefore, perhaps economically rewarding to set up industrial fisheries in those areas where there is a constant supply of suitable fish and where no well-established food fishery already exists. It may be better to earn foreign exchange by exports and to import food that is more acceptable locally; an alternative would be to use locally produced meal to feed animals in the country of production.

Increasing the use of fish

The main technological difficulties in increasing the use of fish for human food are associated with reduction of waste in processing, improving methods of distribution and finding ways of utilising new species, caught on grounds hitherto little exploited.

Where fish can be provided in forms that are traditionally acceptable, or where a fresh or frozen fish industry is developed, there appear to be no institutional difficulties in increasing consumption. Nevertheless the groups known to be most at risk, infants, nursing mothers and the sick, are likely to be neglected unless special feeding programmes are developed for them. This implies state intervention and subsidy. Before feeding programmes are developed, however, surveys need to be carried out to identify the groups most at risk, determine what types of fish or fish products are most acceptable to them, decide how best they can be manufactured and distributed and advise on the best way of subsidising the operation to encourage the development of a permanently viable fish industry.

Where governments intend to encourage fish industries it is necessary to make decisions on the way of utilising the catch. Questions of the availability of capital, of local expertise, the food habits of the population, the reliability of transport systems and the structure of the existing industry must be considered. There are likely to be landings in future of many more small fish and other types of marine animals, such as small crustacea, for the handling and processing of which present techniques are inadequate. FPC is one possible product, but as already indicated its use is limited. Other possibilities involve the use of flesh-stripping machines, of which a number of designs are now available. These will separate flesh from bones, skin, shell, etc. Although the texture of the flesh may be somewhat altered, the technique has considerable potential for the future.

Finally, it should perhaps be stressed that the major problems associated with increasing the use of fish for food are institutional and economic rather than technological. Some technological problems remain to be solved, but a wide range of products exists, some of which are acceptable in most communities. It costs money and requires expertise to develop fisheries and it is in this area that the great difficulties exist.

References

Bardach, J. E., Ryther, J. H. & McLarney, W. O. (1972). *Aquaculture*. Wiley-Interscience, New York.

Beatty, S. A. & Fougère, H. (1957). *The processing of salt fish*. Bulletin of the Fisheries Research Board of Canada No. 112.

Borgstrom, G. (1962). Shellfish protein – nutritive aspects. In: *Fish as food* (ed. G. Borgstrom), vol. 2, pp. 115–47. Academic Press, New York.

Braekkan, O. R. (1962). B-vitamins in fish and shellfish. In: *Fish in nutrition* (ed. E. Heen & R. Kreuzer), pp. 132–40. Fishing News (Books) Ltd., London.

Burgess, G. H. O. (1971). *The alternative uses of fish*. FAO Fishery Report No. 117. FAO, Rome.

Burgess, G. H. O., Cutting, C. L., Lovern, J. A. & Waterman, J. J. (1965). *Fish handling and processing*. HMSO, Edinburgh.

Burgess, G. H. O. & Shewan, J. M. (1970). Intrinsic and extrinsic factors affecting the quality of fish. In: *Proteins as human food* (ed. R. A. Lawrie), pp. 186–99. Butterworths, London.

Connell, J. J. (1970). Properties of fish proteins. In: *Proteins as human food* (ed. R. A. Lawrie), pp. 200–12. Butterworths, London.

Dollar, A. M., Goldner, A. & Olcott, H. S. (1967). Temperature, weight and drip changes during pre-cooking of tuna. *Fishery Industrial Research*, **3**, 19–23.

FAO (1964). Catches and landings 1963. Yearbook of Fishery Statistics No. 16. FAO, Rome.

FAO (1970). *Provisional Indicative World Plan for Agricultural Development. Summary and Main Conclusions*. FAO, Rome.

FAO (1972). Catches and landings 1971. Yearbook of Fishery Statistics No. 32. FAO, Rome.

Graham, H. W. & Edwards, R. L. (1962). The world biomass of marine fishes. In: *Fish in nutrition* (ed. E. Heen & R. Kreuzer), pp. 3–8. Fishing News (Books) Ltd., London.

Gulland, J. A. (1971). *The fish resources of the ocean*. Fishing News (Books) Ltd., London.

Hickling, C. F. (1962). *Fish culture*. Faber & Faber, London.

Higashi, H. (1961). Vitamins in fish. In: *Fish as food* (ed. G. Borgstrom), vol. 2, pp. 411–86. Academic Press, New York.

Higashi, H. (1962). Relationship between processing techniques and the amount of vitamins and minerals in processed fish. In: *Fish in nutrition* (ed. E. Heen & R. Kreuzer), pp. 125–31. Fishing News (Books) Ltd., London.

Knobl, G. (1973). Fish protein concentrates: the state of the art. Reprint from 5th annual offshore technology conference, Houston, Texas.

Lee, D. J. & Putnam, G. B. (1973). The response of rainbow trout to varying protein/energy ratios in a test diet. *Journal of Nutrition*, **103**, 916–22.

Liston, J., Stansby, M. E. & Olcott, H. S. (1963). Bacteriological and chemical basis for deteriorative changes. In: *Industrial fishery technology* (ed. M. E. Stansby), pp. 350–61. Van Nostrand Reinhold, New York.

Liu Chien-kang (1955). On the productivity of two experimental fishponds managed with traditional method of Chinese pisciculture. *Acta Hydrobiologica Sinica* No. 1. (Chinese with English summary.)

Lovern, J. A. (1962). The lipids of fish and changes occurring in them during processing and storage. In: *Fish in nutrition* (ed. E. Heen & R. Kreuzer), pp. 85–111. Fishing News (Books) Ltd., London.

Lovern, J. A. & Wood, H. (1937). Variations in the chemical composition of herring. *Journal of the Marine Biological Association of the United Kingdom*, **22**, 281–93.

Shewan, J. M. (1961). The microbiology of sea water fish. In: *Fish as food* (ed. G. Borgstrom), vol. 1, pp. 487–544. Academic Press, New York.

Stansby, M. E. & Olcott, H. S. (1963). Composition of fish. In: *Industrial fishery technology* (ed. M. E. Stansby), pp. 339–49. Van Nostrand Reinhold, New York.

22. Fungi

W. E. TREVELYAN

When a fungal spore, which is typically a microscopic body only a few micrometres across, is supplied with water and suitable nutrients, it puts out a germ-tube which rapidly extends into a much-branched hyphal thread, the aggregate of such threads in a culture being referred to as mycelium. This filamentous habit of growth differentiates fungi from single-cell organisms such as the yeasts, which otherwise, for example biochemically, they closely resemble. Mycelium may form easily visible mats or strands, often of considerable mechanical strength. Eventually, because of exhaustion of nutrients or for other reasons (Smith & Galbraith, 1971), the mycelium differentiates into special spore-bearing structures or sporophores, which may be microscopic in size (in the case of moulds) or may be quite large (as with mushrooms).

Fungal spores are probably indigestible: but the macroscopic sporophores of certain types of fungi which, as mushrooms and toadstools, have been gathered since time immemorial (Ramsbottom, 1953), and the mycelium of selected fungal strains, have both been advocated as potentially valuable sources of protein for man and for domestic animals (Gray, 1970). For historical and other reasons, the production even on a large scale of sporophores and mycelium involves two quite different technologies. Mushroom cultivation in the Western world (Atkins, 1966; Smith, 1969; Gray, 1970), as in the Orient (Gray, 1970; Smith, 1972) may be considered to be a type of farming operation; but the production of mycelium, which so far has taken place only on a pilot-plant scale, is, or will be, an industrial fermentation process resembling the manufacture of baker's yeast or food yeast. It should be added that huge amounts of fungal mycelium are even now being produced as a by-product of the manufacture, by fermentation, of antibiotics, citric acid, etc. But, according to Hastings (1971), it tends to autolyse 'most unpleasantly', and it usually too rich in phosphates for successful use even as animal feed. It has been reported that mice did not at first take well to a diet containing *Penicillium chrysogenum* mycelium, which had been washed free of penicillin and then dried; but later the animals adapted to it, and gained weight as well as those on a standard diet (Pathak & Seshadri, 1965). In mushroom cultivation, mycelium is a waste product, whereas, in processes for the cultivation of filamentous fungi by industrial fermentation operations, conditions are chosen such that sporulation does not occur.

The common mushroom, or rather a variant of it, *Agaricus bisporus*, was first deliberately cultivated in France, in the seventeenth century. In the temperate regions of Europe and North America, a considerable industry is now supported. At least 70 million kg/yr are produced in the United States (Smith, 1969; Gray, 1970) and some 25 million kg, worth (wholesale) £12 M, in the United Kingdom (Smith, 1969). About 25 kg of mushrooms are needed to provide 1 kg of protein (Hayes, 1969*b*). Hayes (1969*a*) has published a table, reproduced here as Table 22.1, of the world production of mushrooms: he points out that the output of *Agaricus bisporus* is increasing at the average rate of 25 % per annum. UK production units yield annually 70000–80000 kg of protein per hectare, as compared with 80 kg per hectare for beef cattle and 650 for fish farming (Hayes, 1969*b*; cf. Brian, 1972).

Mushroom spawn, i.e., inoculum, is produced in commercial laboratories under strict scientific control. It is propagated by the grower on beds of a composted and pasteurised substrate, based, usually, on a mixture of horse manure and straw. Composting is the secret of success, in that this incomplete microbial degradation brought about by a variety of species of micro-organisms, more especially by thermophilic bacteria, removes easily fermentable carbohydrate, increases organic nitrogen, and in general enables conditions to be established which favour the growth of *Agaricus bisporus* as against possible competitors which, potentially, are faster-growing.

Only a few species of fungi are cultivated by commercial mushroom growers. In the West, *Agaricus bisporus*, as already mentioned: in the Eastern hemisphere, shiitake (*Lentinus edodes*, which grows on dead wood) and the padi-straw mushroom (*Volvariella volvaceae*, which grows, as the common name indicates, on straw). Mushrooms are a fairly good source of protein; Gray (1970) quotes a figure of 27 % of the dry weight for *Agaricus bisporus*, but characterises the protein as incomplete, that is, deficient in one or more essential amino acids. It is an interesting fact that intolerance to mushrooms is very unusual; one case recently recorded (Bergoz, 1971) proved to be due to malabsorption of the sugar trehalose, which is widely found in filamentous fungi (and in yeast). Gilbert & Robinson (1957) remark that there is, apparently, no religious taboo anywhere against eating fungi.

Unfortunately the complexities, not to say hazards, of large-scale cultivation of mushrooms, and also its expense, are such as to make it

Table 22.1. *World production of mushrooms in tonnes × 10^3 wet weight as harvested.* After Hayes (1969*a*)

Country	Year 1950	1960	1966
USA	30.0	50.0	77.4
France	10.0	30.0	49.0
Taiwan	—	1.0	36.8
UK	5.6	17.5	33.8
Holland	0.3	3.0	15.0
Canada	1.5	3.0	7.5
Denmark	0.7	2.5	6.0
W. Germany	0.6	2.5	13.0
Belgium	0.8	2.0	4.0
Switzerland	0.5	1.5	2.8
Sweden	0.4	1.5	2.8
Irish Republic	0.2	0.8	1.6

improbable that they can be developed as a *cheap* protein supplement to the diet, though proposals for their cultivation on various wastes have been made (Gray, 1970). They must at present be regarded as a delicacy, and as a flavouring agent for use by the food manufacturer, who incorporates them into canned or dehydrated soups, sauces, etc.; such uses account for some two-thirds of the United States output (Smith, 1969). With the development, during the Second World War, of large-scale stirred, aerated fermenters which could be operated without risk of microbial contamination, it was natural to consider whether some of the difficulties in mushroom production could not be solved by turning to the propagation of mycelium of the same or related strains of fungi. The considerable amount of research on this topic has been reviewed by Litchfield (1967) and by Worgan (1968).

At least one commercial concern in the USA markets mushroom mycelium (*Morchella* species) made by deep-tank fermentation, but again this finds application in the compounding of soups, etc. (as Morel mushroom flavouring). Litchfield (1968) presents tables showing figures for protein content (total N × 6.25, comprising up to 50 % of the dry weight), and for the content of individual amino acids. The amino acid profiles, like those of micro-organisms in general (Anderson & Jackson, 1958), reveal a deficiency in the sulphur-containing amino acids methionine and cystine, relative to the FAO reference protein. Skinner (1934), working with mould mycelium, was the first to draw attention to this deficiency.

Mycelium produced by industrial fermentation processes often fails to develop the characteristic and delicate flavour of the sporophore. This still leaves such processes open to consideration purely as a means of producing a protein-rich food or feed ingredient, but, in general, other species of fungi – those known collectively as moulds – appear to be better adapted for this purpose. In fact, investigation of the food value of fungal mycelium produced by industrial fermentation processes was developed without reference to the mushroom industry, mainly by German investigators initially, who were impelled to an interest in it by the shortages imposed in both the world wars by blockade.

Originally, it was the production of microbial *fat* to which attention was largely directed (Thatcher, 1954). Lindner (1922), who had devised a process based on the yeast *Trichosporon pullulans* (then known as *Endomyces vernalis*), saw a solution to the problem of making Germany self-sufficient in *protein* in the ammonia–molasses process for cultivating yeast (Mineralhefe) introduced by Delbrück, Hayduck and others. (Both processes were worked out at the Institut für Gärungsgewerbe in Berlin, which was also responsible for the development of *Candida utilis* (torula yeast) production in the nineteen-thirties (Wiley, 1954), and for the suggestion that food yeast might be propagated on hydrocarbons (Just *et al.*, 1951–2).)

Lindner believed that extensive biosynthesis of fat occurred only in surface culture. This notion possibly prompted later investigations on the potentialities of filamentous fungi, since, until the Second World War, moulds were cultivated exclusively in this way. The article by Ward *et al.* (1935) describes the kind of equipment and technique which was commonly used. Partly because of the high labour requirement of such processes they were eventually abandoned in favour of submerged culture, initially introduced as a batch process (Bernhauer *et al.*, 1948; Bernhauer & Rauch, 1948*a*, *b*). In surface or in submerged (batch) culture, a high ratio of assimilable nitrogen to carbon (N/C ratio) was found to favour protein biosynthesis, the reverse the accumulation of fat (Thatcher, 1954). Interest in microbial fat soon waned, and attention became focused on the production of fungal protein.

In 1943, the English author Fawns had very little to say about the potential of filamentous fungi as sources of protein (Fawns, 1943), but in Germany Professor Lembke (as reported by, e.g., Gilbert & Robinson, 1957) had already investigated the effects of eating microfungi, particularly the mycelium of species of *Rhizopus* and of *Fusarium*, both of which were said to be relatively good sources of methionine and cystine. *Fusarium* mycelium, propagated on whey at the East Munich dairy station, showed most promise. However, mould mycelium caused severe diarrhoea unless first completely autolysed by heat treatment.

In view of the fact that the industrial production of yeast had been established and well-understood for a long time (Johnson, 1971), it is pertinent to ask why the large-scale production of filamentous fungi should have received so much attention. The answer probably has to do with the restricted number of compounds which will usefully serve as sources of C for the growth of *Saccharomyces*-type yeast, and its requirement for certain vitamins. (Torula yeast – *Candida utilis* – which is less exacting in both respects, was isolated as late as 1926.) Thus, Pringsheim & Lichtenstein (1920), who investigated the propagation of *Aspergillus fumigatus* on alkali-treated straw, did so because molasses for the cultivation of yeast was no longer available because of the war-time blockade. (Submerged culture of this fungus has recently been investigated by Rogers *et al.*, 1972.) Many strains of fungi will attack starch, a particularly important property because of the widespread distribution of this carbohydrate, and indeed amylolytic enzymes are produced industrially by fermentations involving species of *Aspergillus* or *Rhizopus*.

In considering the development of an industrial fermentation process, one can start with a desired product (e.g., baker's yeast), and look for the most suitable substrate and method of fermentation, or, alternatively, start from the proposition that a substrate, usually some kind of agricultural waste, is available in quantity and look for a suitable *product*, such as a yeast or mould which may serve as a source of edible protein. The second approach has prompted innumerable studies of mould propagation (e.g., that of Church *et al.*, 1973), and is, superficially, very attractive: but, because of costs of collection, stabilisation, and clean-up, wastes are not always so cheap to use as it may at first appear (Litchfield, 1968), and there is always the danger of adsorption of toxic materials from substrates of complex, and sometimes variable, composition (Scrimshaw, 1973).

At present, the process most likely to emerge as the first large-scale commercial production of edible fungal protein is that pioneered by the UK firm of Rank Hovis McDougall (RHM). Here it is interesting to note that, although it is claimed that fungal protein is both desirable in itself and allows cheap, starchy substrates an economically valuable outlet (Spicer, 1971), it is the first consideration which appears to be receiving most emphasis. Already in 1945, Vinson and co-workers claimed that the dried (surface-grown) mycelium of *Fusarium lini* was a better dietary supplement than brewer's yeast. (They were more concerned with B-vitamin content than with protein, though this aspect received some attention, and their interest in *Fusarium* seems originally to have been related to the production of industrial alcohol, cf. Nord & Mull, 1945.)

RHM, as a firm with wide experience of food manufacture, place emphasis on the texture and bland flavour of suitable fungal mycelium, which makes it far more attractive for incorporation into manufactured foodstuffs than yeast. After extensive screening trials, the organism chosen was a strain of *Fusarium graminearum*, which has a remarkably high growth rate: it doubles its mass in 2.4 h at 30 °C (G. L. Solomons, personal communication), thus refuting the assertion (Litchfield, 1968) that fungi invariably grow more slowly than yeasts (Anderson *et al.*, 1973; Solomons, 1973). However, in order to make full use of this property, it is preferable to base the growth medium on sugars; starch must be hydrolysed to glucose, while the preferred substrate for commercial production may be molasses. Some addition of vitamins to the growth medium is also required. Finally, the high growth rate goes hand-in-hand with a high content of nucleic acid, of the same order as that in yeast (purine N about 10 % of total N), and this has to be lowered before the fungal protein can be considered acceptable for human use.

Tables showing a wide range of values for yield and protein content for various types of fungi grown in submerged culture are given by Litchfield (1968) and by Gray (1970). It is doubtful whether these figures are of much interest, since an early stage in the development of any industrial procedure would be screening trials to find an organism which performed satisfactorily in both respects. The analytical problems encountered in trying to arrive at meaningful figures for protein content are summarised by Anderson *et al.* (1973), who also present extensive tables showing the amino acid composition of a number of strains of microfungi. As regards the efficiency of conversion of carbohydrate into fungal protein, it has been shown that, in optimum conditions, yields of a wide variety of micro-organisms grown on a number of different carbonaceous substrates are surprisingly uniform. There are different ways of expressing this regularity; for example Bell (1972) suggests that yield, in terms of grams dry weight, is usually 1.10/g C provided by the substrate. A commercial process for propagating fungi on carbohydrate would in practice aim at a yield of 0.5 g dry weight/g carbohydrate.

The phrase *optimal conditions* implies that all nutrients required (C source, N source, phosphate, metal ions, vitamin-like growth factors) are present in sufficient amount, but it is as important that full aerobiosis is maintained, and that the substrate should be prevented from undergoing secondary reactions. With a substrate such as cellulose, where the hydrolysis of the substrate itself is probably the rate-limiting step of the growth process, there is little danger of diversion of substrate from assimilation into cell substance to the formation of products such as ethanol. With fermentable sugars, though, it is advisable to maintain a low steady-state concentration in the culture which, in commercial yeast manufacture (Burrows, 1970) is done by supplying molasses to an aerated yeast suspension at a rate which is proportional to the amount of yeast present in the fermenter at any stage, and in mould propagation is likely to follow from the use of a continuous culture process (Solomons, 1974).

Even on the laboratory scale, the growth of filamentous fungi by a continuous process is not simple (Rowley & Bull, 1973). A great deal of trouble can be caused by the tendency of filamentous moulds to adhere to surfaces, to block openings, or to foul impellers. It is in fact curious that continuous growth procedures work with moulds as well as they do. Academic studies of the technique have mainly relied on true single-cell organisms such as bacteria, where the number of microbial cells is strictly proportional to the mass of organisms, a consideration which can apply to filamentous fungi only if the mycelium fractures after a certain size has been attained. Genetic factors may be involved here, but it is not without significance that Rowley & Bull recommend very high stirring rates, which suggests that agitation serves a double function of ensuring good aeration and mechanically breaking up mycelium. Questions of power requirements and of damage to the growing mould obviously arise.

As important as the yield per unit weight of substrate is the maximum concentration of microbial mass which can be attained in a fermenter and which is obviously related to the efficient use of fermenter space. In RHM's pilot-plant studies, 30 g dry weight/litre can be attained (G. L. Solomons, personal communication). The rheology of thick suspensions of mycelial strands presents problems, especially in connection with efficient aeration.

In recovery by filtration, or by centrifugation, moulds would appear to hold the advantage over yeast, and even more over bacteria (Spicer, 1971). Whether the drying of mycelium presents any special problems is not known.

A commercially viable unit to produce fungal protein must be large, say 10 to 100000 tonnes/annum. MacLennan *et al.* (1973) think that a minimum size for plants producing microbial protein might be 50000 tonnes/annum. There is, however, still some interest in the possibility of smaller and simpler processes. Thus Reade *et al.* (1972) proposed that fungi should be propagated by a batch process on a slurry of barley supplemented with sources of inorganic nitrogen, the entire fermented product being used as an animal feed. However, the mycelium, when separated from the barley residue, contained a total quantity of protein apparently little more than had been supplied initially in the barley (which had a protein content of 10–12 %), so the degree of enrichment is problematical.

Surface growth on liquid medium is unlikely to be used for the production of fungal protein because the mycelial felt is so inhomogeneous in composition, and may in part consist of autolysed cells. In many tropical countries, however, fermented foods are prepared by the culture of moulds on moist, natural *solid* materials, such as soybeans, or cereals. A good deal of attention has been devoted to such traditional fermented foods (Hesseltine, 1965; Gray, 1970; Pederson, 1971), which succeed in making some forms of protein more attractive to eat (an example of 'biological ennoblement', according to Platt, 1964), even if the actual amount of protein is not increased.

Stanton & Wallbridge (1969) proposed to adapt the attractively simple moist-solids fermentation technique by choosing as the basic substrate material such as cassava flour which was low in protein, and incorporating ammonium salts or urea in the mixture. It was intended that the fermented product should be eaten as such, without further purification. This process resembles the one devised during the First World War by Pringsheim & Lichtenstein: they, however, considered that the residues of nutrient salts in the fermented material made it fit for animal feed only. It is difficult to fault this conclusion. Further study at the Tropical Products Institute of the propagation of *Rhizopus oryzae* on cassava flour which had been moistened with a solution of ammonium phosphate and urea, has shown that growth of the mould ceases when 4.5 g of mycelial protein has been obtained from 100 g dry weight of starting material. There is a concurrent loss of 25 % of the dry weight of the substrate (mainly starch), making the conversion of carbohydrate to protein rather inefficient. Similar figures are cited by Barnes *et al.* (1972), who adapted the procedure to the cultivation of *Sporotrichum thermophile* on moist cellulose (waste paper).

The widespread consumption of mushrooms, tropical mould-fermented foods, cheeses cured with moulds, and the like, has been presented (e.g., by Spicer, 1971) as an argument for the acceptability and safety of mycelial protein in general. Recent studies on the nutritive value, safety and acceptability of microbial protein have been ably reviewed by Kihlberg (1972) and by many others. There seems no reason to doubt that, with careful strain selection and process development, it is possible to produce from molasses, and possibly other carbohydrate sources, together with ammonia, dried fungal protein which has an NPU (net protein utilisation) of about 70 (Spicer, 1971), an acceptable appearance, flavour, and functional properties.

An open question is the effect on man of the 50 % or so of dried mycelium which is *not* protein. Until a standardised product is available and extensive tests have been carried out on it, any pronouncement on this is speculative. What is clear is that the production of such a product under rigid control, and the testing required to establish a reasonable presumption of safety, require resources only likely to be commanded by a big and technologically advanced organisation.

A word should perhaps be said on the subject of mycotoxins. It is undoubtedly rather disturbing to read that *Fusarium graminearum* can produce an oestrogenic mycotoxin (Mirocha *et al.*, 1971), for example. But the use of the same specific name should not lead one to exaggerate the biochemical resemblance between this strain of mould and that chosen by RHM, or the one studied by Vinson *et al.* in 1945. A mycotoxin has been isolated from a strain of *Penicillium roqueforti* (Wei *et al.*, 1973), but connoisseurs of cheese should not necessarily despair. Solomons (1973) examined several hundreds of microfungi, but found only one toxic strain (*Alternaria*).

The part which fungi will play in the future in contributing to the world stock of edible protein is still very difficult to assess. It seems to be taken as certain that a serious shortage of edible protein *will* arise, though little consideration is given to the interaction with other shortages that may also be near. Abundant energy is required to synthesise the ammonia which is so often taken for granted, to operate aeration equipment, agitators, centrifuges, etc. – not to mention transport which is required for distribution. Water of acceptable quality is assumed to be available. In fact no-one can tell how economically attractive, in the distant future, will be the production of the so-called SCP or single-cell protein (though SCEP, single-cell *encapsulated* protein would perhaps be more accurate). It may well be that the greatest contribution made by fungi will be in making available enzymes with the capacity to break down cellulose to more tractable raw materials (Reese *et al.*, 1972).

Meanwhile, mushroom farming will continue to expand its modest, but valuable, contribution to our dietary. Industrially produced mould mycelium will amost certainly make an appearance soon, in a highly industrialised country, dressed up by food technologists as a meat extender. A single factory could produce as much protein in one year as the entire mushroom farming industry of a country. Schemes for the production of cattle food from wastes will continue to be propounded. Progress, in fact, will be made. There will be no short cuts.

References

Anderson, C., Longton, J., Maddix, C., Scammell, G. W. & Solomons, G. L. (1973). The growth of micro-fungi on carbohydrates. Paper presented at 2nd International Symposium on Single-Cell Protein, Massachusetts Institute of Technology, 29–31 May.

Anderson, R. F. & Jackson, R. W. (1958). Essential amino acids in microbial proteins. *Applied Microbiology*, **6**, 369–73.

Atkins, F. C. (1966). *Mushroom growing to-day*, 5th edition. Faber & Faber, London.

Barnes, T. G., Eggins, H. O. W. & Smith, E. L. (1972). Preliminary stages in the development of a process for the microbial upgrading of waste paper. *International Biodeterioration Bulletin*, **8**, 112–16.

Bell, G. H. (1972). Yield factors and their significance. *Process Biochemistry*, **7** (4), 22–5, 34.

Bergoz, R. (1971). Trehalose malabsorption causing intolerance to mushrooms. *Gastroenterology*, **60**, 909–12.

Bernhauer, K., Neithammer, A. & Rauch, J. (1948). Beiträge zur mikrobiologischen Eiweiss- und Fettsynthese. II. Vergleichende Untersuchungen über die Eiweiss- und Fettbildung durch verschiedene Mycelpilze in der Submers-Kultur. *Biochemische Zeitschrift*, **319**, 94–101.

Bernhauer, K. & Rauch, J. (1948*a*). Beiträge zur mikrobiologischen Eiweiss- und Fettsynthese. I. Die grundlegenden Bedingungen für die Eiweiss- und Fettproduktion durch Mycelpilze in der Submers-Kultur. *Biochemische Zeitschrift*, **319**, 77–93.

Bernhauer, K. & Rauch, J. (1948*b*). Beiträge zur mikrobiologischen Eiweiss- und Fettsynthese. III. Zur Methodik der submersen Mycelzüchung in der 'Rührkultur' und deren Anwendung zur Erzeugung von Fettmycel. *Biochemische Zeitschrift*, **319**, 102–19.

Brian, P. W. (1972). The economic value of fungi. *Transactions of the British Mycological Society*, **58**, 359–75.

Burrows, S. (1970). Baker's yeast. In: *The yeasts* (ed. A. H. Rose & J. S. Harrison), vol. 3, *Yeast technology*, pp. 349–420. Academic Press, London.

Church, B. D., Erickson, E. E. & Widmer, C. M. (1973). Fungal digestion of food processing wastes. *Food Technology* (*Chicago*), **27** (2), 36–42.

Fawns, H. T. (1943). Food production by microorganisms. I. Protein production. *Food Manufacture*, **18**, 194–8.

Gilbert, F. A. & Robinson, R. F. (1957). Food from fungi. *Economic Botany*, **11**, 126–45.

Gray, W. D. (1970). *The use of fungi as food and in food processing*. The Chemical Rubber Co., Cleveland, Ohio.

Hastings, J. J. H. (1971). Development of the fermentation industries in Great Britain. *Advances in Applied Microbiology*, **14**, 1–45.

Hayes, W. A. (1969*a*). New techniques with mushroom composts. *Span*, **12**, 162–6.

Hayes, W. A. (1969*b*). Mushrooms, microbes and malnutrition. *New Scientist*, **44**, 450–2.

Hesseltine, C. W. (1965). A millenium of fungi, food and fermentation. *Mycologia*, **57**, 149–97.

Johnson, M. J. (1971). Fermentation – yesterday and tomorrow. *Chemical Technology*, **1**, 338–41.

Just, F., Schnabel, W. & Ullmann, S. (1951–2). Submerse Züchtung von Kohlenwasserstoffzehrenden Hefen und Bakterien. II. *Die Brauerei* (Wissenschaftliche Beilage), **4**, 57–60, 71–5, 100–3; **5**, 8–13.

Kihlberg, R. (1972). The microbe as a source of food. *Annual Review of Microbiology*, **26**, 427–66.

Lindner, P. (1922). Das Problem der biologischen Fettbildung und Fettgewinnung. *Zeitschrift für angewandte Chemie*, **35**, 110–14.

Litchfield, J. H. (1967). Submerged culture of mushroom mycelium. In: *Microbial technology* (ed. H. J. Peppler), pp. 107–44. Van Nostrand Reinhold, New York.

Litchfield, J. H. (1968). The production of fungi. In: *Single-cell protein* (ed. R. I. Mateles & S. R. Tannenbaum), pp. 309–29. MIT Press, Cambridge, Massachusetts.

MacLennan, D. G., Gow, J. S. & Stringer, D. A. (1973). Methanol–bacterium process for SCP. *Process Biochemistry*, **8** (6), 22–4.

Mirocha, C. J., Christensen, C. M. & Nelson, G. H. (1971). F-2 (zearalenone) estrogenic mycotoxin from *Fusarium*. In: *Microbial toxins*, vol. VII, *Algal and fungal toxins* (ed. S. Kadis, A. Ciegler & S. J. Ajl), pp. 107–38. Academic Press, New York.

Nord, F. F. & Mull, R. P. (1945). Recent progress in the biochemistry of *Fusaria*. *Advances in Enzymology*, **5**, 165–205.

Pathak, S. G. & Seshadri, R. (1965). Use of *Penicillium chrysogenum* as animal food. *Applied Microbiology*, **13**, 262–6.

Pederson, C. S. (1971). *Microbiology of food fermentations.* AVI Publishing Co. Inc., Westport, Connecticut.

Platt, B. S. (1964). Biological ennoblement: improvement of the nutritive value of foods and dietary regimes by biological agencies. *Food Technology*, **18**, 662–76.

Pringsheim, H. & Lichtenstein, S. (1920). Versuche zur Anreicherung von Kraftstroh mit Pilzeiweiss. *Cellulosechemie*, **1**, 29–39.

Ramsbottom, J. (1953). *Mushrooms and toadstools.* A study of the activities of fungi. Collins, London.

Reade, A. E., Smith, R. H. & Palmer, R. M. (1972). The production of protein for non-ruminant feeding by growing filamentous fungi on barley. *Biochemical Journal*, **127**, 32P.

Reese, E. T., Mandels, M. & Weiss, A. H. (1972). Cellulose as a novel energy source. *Advances in Biochemical Engineering*, **2**, 181–200.

Rogers, C. J., Coleman, E., Spino, D. F., Purcell, T. C. & Scarpino, P. V. (1972). Production of fungal protein from cellulose and waste cellulosics. *Environmental Science and Technology*, **6**, 715–19.

Rowley, B. I. & Bull, A. T. (1973). Chemostat for the cultivation of moulds. *Laboratory Practice*, **22**, 286–9.

Scrimshaw, N. S. (1973). The future outlook for feeding the human race. The PAG's recommendations Nos. 6 and 7. In: *Proteins from Hydrocarbons* (ed. H. G. de Pontanel), pp. 189–213. Academic Press, London.

Skinner, C. S. (1934). The synthesis of aromatic amino acids from inorganic nitrogen by molds and the value of mold proteins in the diet. *Journal of Bacteriology*, **28**, 95–106.

Smith, J. (1969). Commercial mushroom production. *Process Biochemistry*, **4** (5), 43–6, 52.

Smith, J. (1972). Commercial mushroom production – 2. *Process Biochemistry*, **7** (5), 24–6.

Smith, J. E. & Galbraith, J. C. (1971). Biochemical and physiological aspects of differentiation in the fungi. *Advances in Microbial Physiology*, **5**, 45–134.

Solomons, G. L. (1973). Food from fungi. Microbiology Group symposium, March 1972. *Journal of the Science of Food and Agriculture*, **24**, 637–8.

Solomons, G. L. (1974). Submerged culture production of mycelial biomass. In: *Industrial utilization of mycelial fungi* (ed. J. E. Smith & D. Bury). Edward Arnold, London. (In press.)

Spicer, A. (1971). Protein production by microfungi. *Tropical Science*, **13**, 239–50.

Stanton, W. R. & Wallbridge, A. (1969). Fermented food processes. *Process Biochemistry*, **4** (4), 45–51.

Thatcher, F. S. (1954). Food and feeds from fungi. *Annual Review of Microbiology*, **8**, 449–72.

Vinson, L. J., Cerecedo, L. R., Mull, R. P. & Nord, F. F. (1945). The nutritive value of *Fusaria*. *Science*, **101**, 388–9.

Ward, G. E., Lockwood, L. B., May, O. E. & Herrick, H. T. (1935). Production of fat from glucose by molds. Cultivation of *Penicillium javani-*

cum van Beijma in large-scale laboratory apparatus. *Industrial and Engineering Chemistry*, **27**, 318–22.

Wei, Ru-Dong, Still, P. E., Smalley, E. B., Schnoes, H. K. & Strong, F. M. (1973). Isolation and partial characterization of a mycotoxin from *Penicillium roqueforti. Applied Microbiology*, **25**, 111–14.

Wiley, A. J. (1954). Food and feed yeast. In: *Industrial fermentations* (ed. L. A. Underkofler & R. J. Hickey), vol. I, pp. 307–43. Chemical Publishing Co., New York.

Worgan, J. T. (1968). Culture of the higher fungi. *Progress in Industrial Microbiology*, **8**, 73–139.

23. Yeasts grown on hydrocarbons

C. A. SHACKLADY

The commercial production of yeasts by fermentation on certain hydrocarbons is now an accomplished fact and is taking place in the United Kingdom and France. These yeasts are included in the class of products widely, if somewhat imprecisely, referred to as single-cell proteins (SCP). In contrast to some other materials in this category these yeasts are intended to be used indirectly for human nutrition in that they are currently being produced for inclusion in animal feedstuffs.

Though indirect, their influence on human nutrition may be twofold. In the first place they could facilitate the starting up or extension of animal industries in areas where at present these are rudimentary or non-existent. Secondly they could remove some of the pressure on more conventional animal feed materials, such as oilseeds and fish products, thus releasing more of these products for direct human use.

Processes

These have been described in detail elsewhere (Bennett *et al.*, 1969, 1971); in summary they are as follows. Two alternative feedstocks may be used. One is a mixture of pure linear alkanes (n-paraffins), the other is a middle distillate of oil boiling in the 280–400 °C range. The purity of the n-alkanes must be such as to meet the specification for food grade mineral oil as defined by FDA specification 121.1146 or that recognised by the FAO/WHO. In addition, the n-paraffins should contain less than 1 μg/kg of each of the following polycyclic aromatic compounds; 3,4-benzpyrene, dibenz[*a*, *h*]anthracene, benz[*g*, *h*, *i*]perylene and 3-methylcholanthrene. Such paraffins are usually produced by a molecular sieve process followed by one of several possible finishing treatments.

Both processes are fully continuous. This is a new departure in fermentation technology since, hitherto, the production of yeast on the more conventional carbohydrate substrates has been a batch process.

The degree of purity of the n-paraffins is so high that, after fermentation, the yeast needs only to be concentrated by centrifugation, washed and dried. There is a low level – less than 0.5 % – of residual

Table 23.1. *Characteristics of hydrocarbon-grown yeasts*

% by weight	n-paraffin yeast (G)	middle distillate yeast (L)
Moisture	4–7%	4–7%
Dry Matter Basis		
Crude protein (N × 6.25)	60	68
Total lipids after acid hydrolysis	10	1.5–2.5
Phosphorus	1.6	1.5
Calcium	0.01	0.3
Pepsin digestibility %	> 80	> 80

hydrocarbon associated with the yeast but this has been shown to present no toxic hazard whatsoever, nor would it have been expected to prove an embarrassment in view of the purity of the feedstock.

If middle distillate is used as the feedstock, only a portion – representing the n-paraffin content of the mixed hydrocarbons – is consumed by the micro-organism for growth. The remaining hydrocarbon must be removed by centrifugation, washing and finally by solvent extraction, this last step being unnecessary, as already seen, in the n-paraffin process.

Product

The result, in either case, is a cream-coloured powder with no smell and, virtually, no taste. A typical analysis of the commercially produced material would be as shown in Table 23.1.

It can be seen that the effect of the solvent extraction to which the yeast from middle distillate was subjected has been to remove some of the natural lipids from the yeast and consequently to increase the content of protein.

The mean of a number of amino acid determinations on these yeasts is shown in Table 23.2 and typical figures for extracted soybean meal and fish meal are given for comparison.

Safety of hydrocarbon-grown yeasts

Products which are, in themselves, new or are the result of new technology cannot be assumed to be safe; their freedom from toxicity must be demonstrated by appropriate tests before they are put into general use. For the yeasts with which the author has been associated, a testing programme was devised in collaboration with the Central Institute for Nutrition and Food Research (CIVO) at Zeist, in Holland. This

Table 23.2. *Amino acid composition (g/16 g N)*

Amino acid	Yeast G	Yeast L	Fish meal	Extracted soybean meal
Isoleucine	5.1	5.3	4.6	5.4
Leucine	7.4	7.8	7.3	7.7
Phenylalanine	4.3	4.8	4.0	5.1
Tryosine	3.6	4.0	2.9	2.7
Threonine	4.9	5.4	4.2	4.0
Tryptophan	1.4	1.3	1.2	1.5
Valine	5.9	5.8	5.2	5.0
Arginine	5.1	5.0	5.0	7.7
Histidine	2.1	2.1	2.3	2.4
Lysine	7.4	7.8	7.0	6.5
Cystine	1.1	0.9	1.0	1.4
Methionine	1.8	1.6	2.6	1.4
Total S-containing acids	2.9	2.5	3.6	2.8

programme, which commenced with acute toxicity studies in 1964, is the same in all essentials as that later recommended by the Protein Advisory Group (PAG) of FAO/WHO/UNICEF and first published in 1970 as PAG Guideline No. 6. A summary of the results obtained is presented in Table 23.3.

Nutritional value of hydrocarbon-grown yeasts

When the toxicological experiments made it nearly certain that the yeasts were safe, a series of experiments with poultry, pigs and, later, pre-ruminant calves was started at the Institute for Agricultural Research into Biochemical Products (ILOB) at Wageningen, in Holland in October 1965. Six consecutive generations of poultry and pigs have been produced and this multiple generation study is continuing. In addition, various short and medium term experiments were communicated in a number of papers in the scientific journals and presented at conferences in different parts of the world (Shacklady, 1968, 1969*a*, *b*, 1970, 1972; Shacklady & van der Wal, 1968; van Weerden *et al.*, 1969, 1970, 1972; Gatumel & Shacklady, 1972; Walker, 1972). More recently, workers in research departments in the United Kingdom, France, Germany, Belgium and Italy have published the results of their work on these yeasts none of which conflict with those obtained in the Wageningen experiments.

The main function of the yeasts has been to replace, totally or partially, fish meal and/or soybean meal in pig and poultry rations. In

Table 23.3. *Summary of toxicological experiments with hydrocarbon-grown yeast at CIVO*

Type of experiment	Duration	Animal	Dietary level of yeast	Histopathology	Result
1. Acute toxicity	6 weeks	Rat	40 %	No. Gross pathology only	No differences observed between control and test groups
2. Sub-chronic toxicity	90 days	Rat	10, 20, 30 %	Yes	No evidence of abnormalities related to treatment
3. Chronic toxicity	2 years	Rat	10, 20, 30 %	Yes	As 2
4. Carcinogenicity	2 years	Rat	10, 20, 30 %	Yes	As 2. No evidence of carcinogenic effect
5. Carcinogenicity	1½ years	Mouse	10, 20, 30 %	Yes	As 2. No evidence of carcinogenic effect
6. Multiple generation study	15 generations and continuing	Rat	10, 20, 30 %	Yes	No evidence of teratogenic, mutagenic or other undesirable effects on reproduction
7. Multiple generation study	23 generations and continuing	Japanese quail	10, 20, 30 %	No	No effect on reproduction, fertility or physical characteristics
8. Teratogenicity study	Pregnancy	Rat	10, 20, 30 %	Skeletal abnormalities	No teratogenic effect
9. Mutagenicity study	5-day feeding period for males	Rat	60 %	Skeletal abnormalities	No mutagenic effect
10. Induced toxicity in products from pigs and poultry fed on alkane yeasts	Various	Rat	Various	Gross pathology	No different effects from products of similar animals on yeast-free diets
11. Sensitisation injection† *a.* Intracutaneous *b.* Patch	2 weeks feeding	Rat, guinea pig, pig, monkey	10 %	—	No sensitivity reaction in any of the species

† Yeast from pure n-alkanes only.

pre-ruminant calves they have replaced part of the dried skim milk in liquid feeding.

Van Weerden found the metabolisable energy of Yeast G and Yeast L for 3 and 5 week old chicks to be 12.8 and 10.7 MJ/kg respectively, and for pigs of 35 and 60 kg liveweight to be 16.4 and 14.7 MJ/kg. These conform to results obtained by Shannon & McNab (1973) and by Lewis & Boorman (1971).

Experiments with poultry

Using Yeast L at dietary levels of 10 and 20 % in place of fish and soya meals, the mean percentage egg production over 52 weeks in each of three generations of laying birds was as follows: control 1 (no yeast), 59.2 %; 10 % Yeast L, 59.4 %; control 2 (no yeast), 55.9 %; 20 % Yeast L, 58.6 %.

In the fourth generation, lower levels of total protein in the diet were used and Yeasts G and L formed the only high protein material in the diets. Results over 52 weeks were: control, 54.7 %; 13 % Yeast G, 56.1 %; 12 % Yeast L, 54.9 %.

Shannon & McNab (1972) have published the results of replacing, in two stages, the soybean meal and finally all the soybean and fish meal by 5, 10 and 20 % of Yeast G in a broiler finisher diet. Their results are summarised in Table 23.4.

Experiments with pigs

Up to the fourth generation of pigs at ILOB, 1353 pigs had been produced in 133 litters on the control diet and 1214 from 119 litters on a diet containing 10 % Yeast L, the average size of litter being, therefore, 10.17 for the controls and 10.20 for the yeast groups. Average birth weight was 1315 g for the controls and 1244 g for the yeast groups.

Up to weaning at approximately 25 kg liveweight, the pigs were given either yeast-free diets or diets with 15 % of Yeast L. During this period their mean daily weight gain was 370 g on the control diet and 361 g on the yeast diet. From weaning to slaughter at 110 kg the experimental groups received 0, 7.5 or 15 % of Yeast L and gained an average of 665, 678 and 689 g/day at feed conversion rates of 3.1, 3.04 and 2.99 respectively.

Barber *et al.* (1971) reported that when Yeast G replaced white fish meal in the diets of pigs there was a marginal improvement in both liveweight gain and feed conversion efficiency.

Table 23.4. *Effects of Yeast G on weight gain by poultry*

	Weight gain (g)	kg feed/kg liveweight gain (4–8 weeks)
Control (soya and fish meal)	1043	2.68
5 % Yeast G	1079	2.69
10 % Yeast G	1043	2.58
20 % Yeast G	1081	2.46

Table 23.5. *Effects of Yeast G on weight gain by calves*

	Liveweight gain (kg)	kg feed/kg liveweight gain
Control 0–14 wk	110.5	1.55
10 % Yeast G 0–14 wk	109.0	1.57
10 % Yeast G 0–14 wk	109.3	1.59

Calves

For the pre-ruminant calf on a liquid diet, approximately 2 parts by weight of dried skim milk may be replaced by 1 part by weight of hydrocarbon grown yeast and 1 part by weight of dried whey. A typical experiment at ILOB gave the result with two commercial samples of Yeast G as shown in Table 23.5.

Other types of livestock

Although not investigated to date as extensively as pigs, poultry and calves, very promising results are being obtained in the use of hydrocarbon-grown yeasts for lambs, rabbits and – especially – in fish farming. Indeed there does not appear to be any species of domesticated animal for which these yeasts are not entirely suitable.

Summary and conclusion

The results of the extensive toxicological and nutritional experiments carried out on the yeasts made by the processes outlined here demonstrate conclusively their safety and nutritive value. It must be emphasised however that materials made by other processes must be shown to be equally safe and their nutritional value demonstrated before they too can be accepted for use in animal feeds.

It may be evident from what has been said that the production of hydrocarbon-grown yeasts on a scale large enough to be commercially viable requires an animal feed market of substantial size and sufficiently well organised to absorb it. Furthermore it requires a high capital investment but does not employ much labour. Whilst this is an attraction in some countries, it may not be so universally. On the other hand, points in favour of hydrocarbon-grown yeasts are that they:

(i) will be produced where they are needed and may thus lighten the burden on imports;
(ii) will be more constant in composition than most products derived from conventional agricultural methods;
(iii) have a long shelf life;
(iv) may be expected to be more stable in price than agricultural commodities;
(v) will not be subjected to variations in seasonal or climatic conditions such as droughts, floods etc.;
(vi) will have a nutritional value at least equal to that of analogous products derived from macro-agriculture;
(vii) will make no demands on land or other facilities required for existing agricultural purposes.

These yeasts do not represent any threat to existing agricultural products; they are complementary to them and add to the pool of raw materials from which the quantity and quality of livestock may be improved, to the ultimate benefit of all.

Permission to publish this paper has been given by the British Petroleum Company Limited.

References

Barber, R. S., Braude, R., Mitchell, K. G. & Myers, A. W. (1971). *British Journal of Nutrition*, **25**, 285–94.

Bennett, I. C., Hondermarck, J. C. & Todd, J. R. (1969). *Hydrocarbon Processing*, March 1969.

Bennett, I. C., Yeo, A. A. & Gosling, J. A. (1971). *Chemical Engineering*, 27 Dec. 1971, 45–7.

Gatumel, E. & Shacklady, C. A. (1972). In: Proceedings of the 2nd World Congress on Animal Feeding, pp. 417–46.

Lewis, D. & Boorman, N. (1971). Paper given at Proceedings of the Xth International Congress on Animal Production, Versailles.

PAG (1970). Protein Advisory Group Guideline No. 6. United Nations Organisation, New York.

Shacklady, C. A. (1968). *Proceedings of the Nutrition Society*, **28**, 91–7.

Shacklady, C. A. (1969*a*). *Biotechnology and Bioengineering Symposium*, **1**, 77–97.
Shacklady, C. A. (1969*b*). *Voeding*, **30**, 574–9.
Shacklady, C. A. (1970). In: Proceedings of the 3rd International Congress on Food Science and Technology, pp. 743–7.
Shacklady, C. A. (1972). *World Review of Nutrition and Dietetics*, **14**, 154–79.
Shacklady, C. A. & van der Wal, P. (1968). Paper given at 2nd World Conference on Animal Production, Maryland.
Shannon, D. W. F. & McNab, J. M. (1972). *British Poultry Science*, **13**, 267–72.
Shannon, D. W. F. & McNab, J. M. (1973). *Journal of the Science of Food and Agriculture*, **24**, 27–34.
van Weerden, E. J., Shacklady, C. A. & van der Wal, P. (1969). Paper given at 8th International Congress of Nutrition, Prague.
van Weerden, E. J., Shacklady, C. A. & van der Wal, P. (1970). *British Poultry Science*, **11**, 189–95.
van Weerden, E. J., Shacklady, C. A., van der Wal, P. & Schutte, J. B. (1972). *Archiv für Geflügelkunde*.
Walker, T. (1972). *Nutrition and Food Science*, April 1972.

24. Variation in the composition of bacteria and yeast and its significance to single-cell protein production*

C. L. COONEY & S. R. TANNENBAUM

The case for producing microbial cells as a source of protein and other nutrients is based primarily upon the expected economy and scale of production, compared to competing sources such as fish meal, oilseed meals, leaf protein concentrates, etc. Articles and books on the subject of single-cell protein (SCP) have dealt with the numerous and complicated factors (Table 24.1) associated with the ultimate choice of an organism or substrate (Mateles & Tannenbaum, 1968). This article will consider certain factors which influence the nutritive value of bacteria and yeast, specifically factors related to cell composition and variation of cell composition.

Bacteria and yeast are complex in structure and nitrogen (N) is distributed in many types of chemical structures, including proteins, amino acids, peptides, nucleic acids and amino sugars. The location of the constituents among the cell fractions is summarized in Table 24.2. Some of the amino acids of the cell may be contained in the cell wall, in peptide or non-peptide linkages which are more resistant to hydrolysis by mammalian digestive enzymes than plant or animal proteins. The amino acid composition of some cell fractions, such as the cell wall, is much simpler than that of complete proteins. A typical example is given in Table 24.3 for *L. casei* (Ikawa & Snell, 1956). Separation of the cell wall from the remainder of the cell protein may have some influence on the nutritive value of the proteins, but the overall effect is not expected to be large (Tannenbaum & Miller, 1967). In yeast and gram-positive bacteria there appears to be a total lack of sulfur and aromatic amino acids in the cell wall.

The essential amino acid content of micro-organisms is probably the best indicator of the overall nutritional utility of the cell for animals and man. In this respect micro-organisms are not as rich a source of

* Publication Number 2319 from the Department of Nutrition and Food Science, Massachusetts Institute of Technology, Cambridge, Massachusetts 02139.

Table 24.1. *Factors associated with the choice of an organism for SCP*

Safety	Type of process
Nutritive value	Yield factors for organism
Palatability	Energy economy: heating, cooling, etc.
Cost of substrates	Plant location

Table 24.2. *Distribution of chemical constituents in microbial cells*

Cell fraction	Major components
Capsule	Complex polysaccharide, polypeptides
Cell wall	Lipopolysaccharide, polypeptides, amino sugars
Cytoplasm	
Cytoplasmic membrane	Lipoprotein, RNA
Ribosomal fraction	Protein, RNA
Nuclear bodies	Protein, RNA, DNA
Mitochondria	Lipid, protein
Cell sap	Lipid, protein, polysaccharide, small organic molecules

Table 24.3. *Amino acid composition of isolated cell walls of* Lactobacillus casei†

Amino acid	g/100 g cell wall	Amino acid	g/100 g cell wall
Arginine	—	Valine	1.4
Histidine	—	Aspartic acid	3.1
Lysine	2.3	Glutamic acid	7.2
Leucine	1.4	Serine	0.6
Isoleucine	1.4	Proline	—
Methionine	—	Glycine	1.0
Phenylalanine	—	Alanine	8.4
Tyrosine	—	Diaminopimelic acid	6.7

† Ikawa & Snell (1956).

high quality protein as animal muscle, and most microbial species are deficient in the sulfur amino acids. Johnson (1967) compared the essential amino acid contents of a number of micro-organisms, and his results are presented in Tables 24.4 and 24.5. In Table 24.4 the choice of essential amino acids is shown. Aside from the basic amino acids lysine and histidine, each amino acid has a rather narrow concentration range for these microbial samples. As can be seen from Table 24.5, the differences between bacteria and yeast are not large. Although bacteria occupy the higher end of the scale, there is considerable overlap.

Table 24.4. *Average amino acid distribution of 11 microbial samples with standard deviation from average*†

Amino acid	% total essential amino acids	Standard deviation
Histidine	7.22	1.05
Arginine	11.18	0.36
Lysine	15.31	1.01
Leucine	16.57	0.53
Isoleucine	11.34	0.57
Valine	11.85	0.37
Methionine	3.81	0.31
Threonine	10.23	0.33
Phenylalanine	9.43	0.67

† Johnson (1967).

Table 24.5. *Essential amino acid contents of 11 microbial samples*†

Sample	Content (dry wt)	
	Essential amino acid	Nitrogen
	Bacteria	
Staphylococcus aureus	21.6	10.75
Escherichia coli	33.1	13.19
Bacillus subtilis	23.8	10.07
	Yeasts	
Saccharomyces cerevisiae (av.)	17.1–23.8	5.9–8.2
Saccharomyces cerevisiae	23.1	8.94
Torula yeast	29.5	8.35
Torula yeast	24.4	7.47
	Molds	
Aspergillus niger	9.2	5.21
Penicillium notatum	12.8	6.13
Rhizopus nigricana	9.6	5.80
	Mushroom	
Tricholoma nudum	20.8	8.64
	Non-microbial samples	
Animal muscle (av.)	48.1	15.4
Fish meal (av.)	32.1	9.8
Alfalfa meal	6.9	2.72

† Johnson (1967).

Environmental effects on cell composition

The nutritional value of the cell protein is determined by its amino acid profile; however, if whole cells or protein isolates are to be used for humans, it will be necessary to minimize the nucleic acid content of the crude protein source since the amount of nucleic acid which can be metabolized per day is limited (Waslien *et al.*, 1970). From the previous discussion and data in the literature (Herbert, 1961; Sykes & Tempest, 1965; Brown & Rose, 1969) it is also clear that the macromolecular composition of micro-organisms varies. Furthermore, for a given micro-organism, cell macromolecular composition is dependent on the rate and temperature at which the cells are grown.

In this discussion, the important markers of cell composition are the concentration of protein and RNA as a percentage of cell dry weight and the ratio of protein to RNA. The concentration of protein per cell is dependent on the type of micro-organism but is fairly constant for that organism under a wide range of environmental conditions. When protein content is expressed as percentage of dry cell weight, it is found to vary inversely with the concentration of other cell components. On the other hand, since most of the RNA is used to synthesize protein, the protein to RNA ratio is proportional to the rate of cell growth. The important question which arises with regard to single-cell protein production is: what is the extent to which we can control and manipulate cell composition in order to optimize the protein concentration and protein to RNA ratio? Previous work using bacteria (Tempest & Dicks, 1967) and yeast (Brown & Rose, 1969) has shown that protein concentration is only slightly dependent on growth rate and temperature while protein to RNA ratios decrease with increasing growth rate and with decreasing temperatures. Also, data of Tempest & Dicks (1967) suggest that the macromolecular composition is dependent on the nature of the growth-limiting nutrient for cells growing in a chemostat.

While it was clear in 1967 that cell composition could be manipulated by control of the growth environment, all of these studies in continuous culture employed single growth-limiting nutrients with all other essential nutrients available to the cell in excess. The question then arose, could even greater control over cell composition be obtained by using dual-nutrient limitations?

Based on available information, Cooney (1970) hypothesized that a growth limitation by C would restrict the cell's energy supply or supply of carbon skeletons for biosynthesis, a growth limitation by N or sulfate

Table 24.6. *Macromolecular composition of* A. aerogenes *grown in carbon-, nitrogen-, and phosphate-limited chemostats*

Limiting nutrient (mg/l)	Inlet glucose (500 mg/l)			Inlet ammonia (40 mg/l)			Inlet phosphorus (7.7 mg/l)		
Dilution rate (h^{-1})	0.22	0.45	0.77	0.24	0.50	0.75	0.25	0.50	0.70
Protein (%)	69.3	63.6	72	52.8	52.4	58.0	72.0	67.0	55.5
RNA (%)	13.3	18.2	24.9	10.5	17.0	20.9	10.2	17.1	19.2
DNA (%)	5.8	5.1	1.9	4.5	1.7	2.5	3.6	3.5	1.7
Carbohydrate (%)	5.0	6.5	8.7	19.1	9.7	9.5	16.9	6.3	5.7
$\frac{\text{Protein}}{\text{RNA}}$	5.3	3.5	2.9	5.0	3.1	2.8	7.1	3.9	2.9
$\frac{\text{Protein}}{\text{RNA}}$ × dilution rate	1.2	1.6	2.2	1.2	1.6	2.1	1.8	1.9	2.0

would restrict protein synthesis and a growth limitation by phosphate, potassium, or magnesium would restrict nucleic acid synthesis. This hypothesis then provided a basis for selecting pairs of limiting nutrients which would allow one to attempt to control selectively the synthesis of various cell components. Using *Aerobacter aerogenes* as a model system for these studies, Cooney & Mateles (1971) examined the limits to which one could manipulate cell composition by the imposition of dual-nutrient limitations. From studies using C and/or N and N and/or phosphate limitation, their results showed that the protein content of the bacteria could be varied from 52 to 72 % and the RNA content from 10 to 25 %. The protein to RNA ratio at a given growth rate, however, was the same in C- and/or N-limited chemostats despite variations in the absolute concentration of protein and RNA. On the other hand, the protein to RNA ratio under phosphate-limited conditions was increased by as much as 40 % at the lowest specific growth rates employed (0.25 h^{-1}). The difference between phosphate-limited growth and the other limitations decreased to zero as the growth rate was increased to 75 % of the maximum. These results are consistent with the hypothesis that a phosphate limitation acts to restrict nucleic acid synthesis while not restricting protein synthesis. A summary of the effect of single-nutrient limitation with the C, N, and phosphorus source is given in Table 24.6. When two nutrients are used simultaneously, the cell composition data fall in between the extremes shown in Table 24.6 (Cooney, 1970); thus cell composition appears to be continually variable between the extremes.

Having now examined the effect of growth environment on the protein content of micro-organisms, it is appropriate to consider protein quality. The early literature on the effects of environmental conditions on amino acid profiles of micro-organisms is limited and difficult to interpret. The primary reason for the difficulty is that investigators generally used whole cell preparations containing free amino acids and peptides, and pools of amino acids are known to vary with the growth environment (Tempest *et al.*, 1970). In the production of SCP, these small molecular weight components could be lost in processing, hence an amino acid profile including these compounds might not truly reflect the quality of the final product.

Recently, Alroy & Tannenbaum (1973) examined the extent to which one can manipulate the amino acid profile of micro-organisms by control of the growth environment. In this work the yeast *Candida utilis* was grown in chemostat culture and the effect of temperature (15 to 37.5 °C), pH (3.0 to 7.5), specific growth rate (0.06 to 0.42 h^{-1}) and N source (NH_4^+ and NO_3^-) on its amino acid profile were examined. Cells were extracted with hot trichloroacetic acid to remove small molecular weight components.

Results from these studies show that while the profile of whole unextracted cells varies considerably (e.g., arginine varies twofold over a range of growth rates), the profiles of extracted cells exhibit only relatively minor variations. The most significant changes are an increase in lysine and arginine (12 and 10 % respectively) and a decrease in serine (24 %) with an increase in specific growth rate from 0.06 to 0.42 h^{-1} as shown by the data in Table 24.7 (Alroy, 1971). These changes coincide with the fact that ribosomal proteins are relatively rich in lysine and arginine and poor in serine and that the cellular content of ribosomes increases with increasing growth rate. Similar results were also found by Alroy (1971) in studies using the bacterium *A. aerogenes*. Because growth temperature affects the ratio of protein to RNA, amino acid profiles also vary with temperature.

As temperature is increased the protein to nucleic acid ratio increases due to a decreased amount of RNA. Interestingly, if cell composition data are examined as a function of the ratio of the growth rate to the maximum growth rate at a given temperature, then the data fall on a single curve. This allows one to predict cell composition over a range of both growth rates and temperatures. Having experimentally measured the variation in amino acid profiles, it is interesting to examine the problem mathematically to see the extent to which amino acid profiles could be expected to vary. From mass balances on an amino acid in given protein species and in the total cell protein, it is possible to calcu-

Table 24.7. *Amino acid composition (% mole of total amino acids) of* C. utilis *at various dilution rates.* (Nitrogen source = NH_4^+, D_m‡ = 0.45 h^{-1}, pH = 5.9, T = 30 °C.)

D† (h^{-1})	0.06		0.12		0.23		0.35	
D/D_m	0.13		0.26		0.51		0.78	
No. of analyses	2		2		2		2	
Lysine	5.76	±0.49	5.79	±0.18	6.89	±0.20	7.28	±0.62
Histidine	1.48	±0.09	1.60	±0.05	1.73	±0.13	1.67	±0.16
Arginine	3.33	±0.31	3.85	±0.13	5.03	±0.18	6.10	±0.64
Aspartic acid	8.91	±0.18	8.61	±0.19	8.37	±0.24	9.03	±0.77
Threonine	6.78	±0.06	7.15	±0.19	6.80	±0.18	6.73	±0.49
Serine	7.75	±0.14	7.30	±0.02	7.36	±0.24	6.86	±0.52
Glutamic acid	20.35	±0.00	18.77	±0.01	19.30	±0.13	18.78	±0.87
Proline	7.17	±1.0	3.70	±0.19	3.38	±0.05	3.72	±0.01
Glycine	7.80	±0.15	7.43	±0.54	7.65	±0.05	7.58	±0.20
Alanine	9.25	±0.19	10.49	±0.46	9.44	±0.04	9.20	±0.34
Valine	6.25	±0.23	6.33	±0.21	6.20	±0.23	6.14	±0.59
1/2 Cystine	0.79	±0.86	1.07	±0.11	0.80	±0.71	0.66	±0.71
Methionine	0.76	±0.09	0.91	±0.01	0.80	±0.18	0.73	±0.25
Isoleucine	4.25	±0.06	4.27	±0.06	4.14	±0.03	3.97	±0.41
Leucine	6.59	±0.11	6.75	±0.20	6.47	±0.06	6.25	±0.75
Tyrosine	2.63	±0.06	2.78	±0.11	2.61	±0.06	2.45	±0.46
Phenylalanine	3.12	±0.06	3.17	±0.08	3.01	±0.06	2.82	±0.40

† D, the steady state dilution rate in continuous culture. ‡ D_m, the maximum dilution rate as obtained by the wash-out technique.

late the change in concentration of the protein species that would be required to change significantly the total concentration of the amino acid in the cell. The results of such a theoretical analysis show that it requires very large changes in a protein species rich in an amino acid to alter the acid profile. This is in complete agreement with the experimentally observed results that the profile is relatively invariable except when a large change in the concentration of a specific class of proteins occurs, e.g., ribosomal proteins which are rich in lysine and arginine.

In summary, these studies on the effects of growth environment on cell macromolecular composition and amino acid profiles have defined the extent to which one can manipulate these parameters of SCP quality through environmental control. The protein to RNA ratio is shown to be, with one exception, a unique function of the ratio of growth rate to the maximum growth rate. The one exception occurs when cells are grown at less than 75 % of the maximum rate under conditions of phosphate limitation which is hypothesized to restrict nucleic acid synthesis. While the ability to alter amino acid profiles through environmental control is limited, it is clear that some variation does occur and that the results are in agreement with those theoretically predicted.

References

Alroy, Y. (1971). Variations in macromolecular and amino acid composition of microorganisms with environmental conditions. PhD Thesis. Massachusetts Institute of Technology, Cambridge, Massachusetts.

Alroy, Y. & Tannenbaum, S. R. (1973). The influence of environmental conditions on the macromolecular composition of *Candida utilis*. *Biotechnology and Bioengineering*, **15**, 239–56.

Brown, C. M. & Rose, A. H. (1969). Effects of temperature on composition and cell volume of *Candida utilis*. *Journal of Bacteriology*, **97**, 261–72.

Cooney, C. L. (1970). Double nutritional deficiencies in continuous microbial culture. PhD Thesis. Massachusetts Institute of Technology, Cambridge, Massachusetts.

Cooney, C. L. & Mateles, R. I. (1971). Fermentation kinetics. Recent advances in microbiology. International congress for microbiology, Mexico City, pp. 441–9.

Herbert, D. (1961). The chemical composition of microorganisms as a function of their environment. In: *Microbial reaction to environment* (ed. G. G. Meynell & H. Gooder). *Symposia of the Society for General Microbiology*, **11**, 391–416. Cambridge University Press, London.

Ikawa, M. & Snell, E. E. (1956). *Biochimica et Biophysica Acta*, **19**, 576.

Johnson, M. (1967). Growth of microbial cells on hydrocarbons. *Science*, **155**, 1515.

Mateles, R. I. & Tannenbaum, S. R. (eds.) (1968). *Single-cell protein*. MIT Press, Cambridge, Massachusetts.

Sykes, J. & Tempest, D. W. (1965). The effect of magnesium and of carbon limitation on the macromolecular organization and metabolic activity of *Pseudomonas* sp., strain C-IB. *Biochimica et Biophysica Acta*, **103**, 93–108.

Tannenbaum, S. R. & Miller, S. A. (1967). Effect of cell fragmentation on nutritive value of *Bacillus megaterium*. *Nature, London*, **214**, 1261.

Tempest, D. W. & Dicks, J. W. (1967). Interrelationships between potassium, magnesium, phosphorus and ribonucleic acid in the growth of *Aerobacter aerogenes* in a chemostat. In: *Microbial physiology and continuous culture* (ed. E. O. Powell *et al.*). Her Majesty's Stationery Office, London.

Tempest, D. W., Meers, J. I. & Brown, C. M. (1970). Influence of environment on the content and composition of microbial free amino acid pools. *Journal of General Microbiology*, **64**, 171–85.

Waslien, C. I., Calloway, D. H., Margen, S. & Costa, F. (1970). Uric acid levels in men fed algae and yeast as protein sources. *Journal of Food Science*, **35**, 294–8.

PART IV

The use of novel foods

25. Quality standards, safety and legislation

F. AYLWARD

The primary reason for food regulations is, or should be, the *protection* of the *consumer* or of the *purchaser*, but there may also be legitimate subsidiary reasons such as the promotion of international trade. Consumer protection (see Table 25.1) may be considered in terms of four overlapping concepts: (i) the prevention of fraudulent practices; (ii) health safety in terms of microbiology and hygiene; (iii) health safety in terms of freedom from toxic substances; and (iv) nutritional value.

Much of the early impetus for regulations came from the desire of local or national governments to remedy abuses and to protect the public from frauds, such as falsifying weights, or adulteration of foods, including adulterations designed to conceal defects and in particular to mask taints arising from decomposition. In the post-Pasteur period, from 1870 onwards, increasing attention was given to hygienic standards, using techniques for the control of micro-organisms; and other tests were introduced later to monitor contamination from other biological sources. In recent years increasing attention has been given to the hazards to the consumer arising from toxic chemicals in three broad groups: those arising from biological action (for example the aflatoxins produced by mould action); those coming from the environment (e.g., pesticide residues, and heavy metals in fish contaminated by industrial effluents); and those resulting from the use of processing aids (e.g., chemical preservatives and emulsifiers). The final section of Table 25.1 notes the importance of the quality of the initial 'freshly harvested' food materials and of maintaining quality from farm (or other sources) to consumer.

Regulatory and advisory bodies

Food regulations were introduced in some European countries many centuries ago (rules governing the sale of bread in England were promulgated in the Middle Ages) but food legislation in its present form is a product of the past century. Over the past hundred years most countries in Europe, and many elsewhere, have adopted regulations on a local, regional or national basis.

Table 25.1. *Consumer protection*

Objective	Examples
(i) Prevention of fraudulent practices	(*a*) falsifying weights (*b*) additions and dilution (*c*) substitutions (*d*) concealment of contamination
(ii) Health safety in terms of microbiology, pest-control and hygiene	(*a*) regulations fixing maximum counts for specific micro-organisms (*b*) mould and other counts (*c*) filth tests
(iii) Health safety in relation to chemical agents	(*a*) toxicity from biological action (*b*) accidental contamination of foods from environment (*c*) use of chemical additives
(iv) Maintenance of nutritional value	(*a*) quality of raw materials (*b*) maintenance of standards during storage and processing

Note: There are inter-connections and overlapping between (i), (ii), (iii) and (iv).

The types of regulations vary with the country; they range from legislation passed by national parliaments, to voluntary *codes of practice* accepted by trade associations and other groups; in countries with a federal constitution (for example, in the United States or West Germany) a distinction must be made between state and federal action. The implementation of regulations may be the responsibility of an inspectorate organised through a national body; in the United Kingdom and some other countries local government agencies play an important role.

International trade in foodstuffs can lead to conflicts of interest in respect to regulations and in recent years various regional bodies have come into being; thus the member countries of the European Economic Community have established machinery for the *harmonisation* of food law. On the international plane, the International Standards Organisation (ISO) is concerned with the quality appraisal methods and standards for many products including some agricultural produce and foodstuffs. In respect to foods, a central place is occupied by the Agencies of the United Nations and in particular by the Food and Agriculture Organisation (FAO) and the World Health Organisation (WHO). These bodies, in co-operation with national member governments, have sponsored the *Codex Alimentarius* which through its various committees is undertaking a systematic examination of standards, and is dealing in detail with questions of food hygiene, safety and nutritional value. Much of the work of the *Codex Alimentarius* committees has been concerned up to now with traditional commodities

Table 25.2. *Protein sources*

Source	Type	Examples
From agriculture or equivalent	Oilseeds	Soybean Groundnut (peanut) Rapeseed Cotton seed Sesame seed Sunflower seed
	Legumes (other than oilseeds)	Broad bean (*Vicia faba*)
	Leaf protein	From many types of leaves
	Algae	*Spirulina* species
From fisheries	Fish protein	From many types of fish
From biosyntheses using micro-organisms	Carbohydrate as carbon source Hydrocarbons (e.g., from petroleum) as carbon source	Products using different types of bacteria or fungi

(e.g., cereals, milk products, canned goods), but another part of the UN 'family', namely the Protein Advisory Group (PAG) has devoted special attention to new protein foods. It was in fact established in 1952 as a specialist body to advise the FAO and WHO and also the United Nations Children's Fund (UNICEF) on protein products, with special reference to their use by 'vulnerable' groups in low-income countries. PAG is not, of course, a *legislative* body but it has produced a series of Guidelines and policy papers which will be referred to later.

Although much of the earlier work on novel protein sources was directed towards protein and food supplies in *low-income* countries, it is in *industrialised* countries – and in particular in North America – that most progress has been made in producing and using newer sources of protein for human and/or farm animal use. I propose, therefore, to examine the problems that have been encountered in the use of new proteins in the United States and in some other industrialised countries and to leave over to a later section problems in low-income regions. In considering the protein sources, distinction must be made between proteins from agriculture and fisheries, and proteins from fermentations (Table 25.2).

In practice, regulations may differ depending on the end use of the protein; for example, whether it is to be used primarily as a component of other foods, as a food in its own right, or as a major component of foods for babies and infants.

Table 25.3. *Production of toxic materials and/or decline in nutritional value during storage or processing of oilseeds*

Action	Prevention
Storage	
Aflatoxin formation arising from mould growth	Control and improvement of storage conditions
Processing	
Solvent residues from extraction Products of chemical interaction of solvents and components of seed	Careful choice of solvents with temperature control
Accidental contamination, e.g., heavy metals	Good housekeeping
Decline in nutritive value of protein; e.g., changes resulting from high temperature and other conditions	Control of temperature of extraction

Proteins from agriculture and fisheries

Oilseeds

Most oilseeds contain protein of good nutritional value, but for any one seed there may be a wide variation depending on varietal differences and methods of processing. Variations also arise from the presence of toxic or potentially toxic material in the seeds, and from additives or from changes produced in processing. Quality standards have to be defined, therefore, for each individual oilseed, taking into account the naturally occurring toxicants that may be present.

Several reviews and monographs detail the wide range of minor components in legumes and other crops, and dispose of the concept that *natural* materials are inherently good. Hence the necessity of devising extraction and fractionation procedures to secure the elimination of toxins or anti-nutrients. Some other potential hazards from oilseeds are listed in Table 25.3; they range from the presence of aflatoxin in groundnuts and other products, through problems from certain solvents, to the effects of excessive heat treatment on protein quality.

Soya products in fermented and other forms have been for centuries an important part of the diet of people in China and other parts of Asia. Peanuts have long been accepted as a foodstuff and one product – namely peanut butter – has been a common food item in the United States for the past fifty years. Soya preparations have been used as an additive to flour for bread-making in many countries for the past thirty years. In the United States, and to a lesser extent in other countries, there has been intensive research and development work on soya protein preparations, including isolates and textured products. There is general agreement that oilseed proteins can make an important contribution to the human diet providing (*a*) they meet quality specifications regarding maximum amounts of certain minor components and (*b*) that when used as a major food (as distinct from a minor ingredient) information should be forthcoming on the nutritional value in terms of protein and other nutrients. Thus, with the agreement of the US Food and Drug Administration, oilseed proteins are now accepted in school meal programmes (up to 25 % textured vegetable protein may be incorporated in hamburgers). This is an important development; it will enable these and parallel institutional programmes to proceed in spite of the rising costs of meat and other animal products.

The specifications proposed in regulations vary somewhat with the protein source; with cottonseed, for example, the content of gossypol in the final product must be kept below defined limits. With some other protein sources, further experimental work is still required – see, for example, current investigations on rapeseed and other members of the *Brassica* family. The US regulations cover questions of nutritional value of proteins and propose fortification where necessary. These regulations are modelled to some extent on the principles worked out for the fortification of margarine considered as a butter substitute.

Other produce of agriculture

Work on leaf protein for animal feed has expanded rapidly over the past five years with the commercial production in California of low-fibre high protein material from alfalfa and other sources, and other commercial activities in Sweden and elsewhere. There has also been a growth of interest and experimentation directed towards its direct human use. It appears that few toxicity problems will arise providing the choice of crops excludes those which are known to be toxic. In terms of nutritional standards, it is now accepted that leaf protein of high biological value can be produced; guidelines are required to cover processing methods and storage. Recent research has indicated the potential value of the non-protein constituents, and in particular lipids, in leaf preparations.

Other agricultural sources of proteins include cereals and legumes (i.e., apart from oilseeds). The broad bean (*Vicia faba*) is potentially an important source in many countries; it is known, however, that factors are present which can promote the onset of the disease favism in individuals who have certain genetic deficiences. The nature and activities of these factors are now being investigated in a number of laboratories with the encouragement of WHO, FAO and other bodies. Clearly, quality standards for *Vicia faba* preparations should include guidance on the levels of these factors; hence the importance of identifying the factors and of devising appropriate methods of assay.

Fisheries

The financial investment in research and development work on fish protein concentrates is probably second only to soya. Quality standards have been defined in the United States and preparations have been used in human food. There is now, however, a widespread belief that concentrates of the type employed are unlikely to make any large contribution to food supplies in developing countries.

Proteins from fermentations (single-cell protein)

For animal feed

An increasing body of literature on single-cell protein suggests there is little doubt about its potential contribution to animal feed. It has, in fact, been used for this purpose in various parts of western Europe and also in central and eastern Europe on an experimental scale; full commercial use is planned in several countries. Published results demonstrate the potential nutritional value of protein obtained by different methods and indicate that, with careful control of production and extraction procedures, animal feed of good quality can be prepared (Table 25.4).

The control of animal feeding stuffs is, in most countries, limited; questions are now being raised in the United States and elsewhere about the effects of feed components on the quality of carcass meat or other products, in particular milk and eggs. In Japan, a pioneer in the development of single-cell protein, a hold-up has occurred in the production and use for animal feed of proteins produced on a hydrocarbon substrate. On the other hand, it is understood that considerable developments have taken place in the USSR and that safety guidelines have been established.

Table 25.4. *Quality appraisal of proteins from biosynthesis*

Possible sources of toxicity	Action and control	
From raw materials		
Hydrocarbons – possible occurrence of branched chain and polynuclear aromatics	Careful choice of raw materials	Selected chemical (and when appropriate) biological tests on all batches
Cellulose and other wastes may contain toxic materials including heavy metals	Extraction procedures on harvested products	
From fermentations		
Nucleic acids	Removal of nucleic acids by enzyme or extraction methods	Selected tests for non-protein nitrogen and for nucleic acids and/or components
Other toxic components	Careful choice of micro-organisms and of fermentation conditions	Selected tests on harvested material and on final products

For direct human use

So far as is known, no legislative or regulatory body in any country has so far sanctioned the direct human use of proteins from fermentations apart from the use of yeast cultivated by traditional procedures. The subject must therefore be regarded as *sub judice*, but some general comments may be made. It seems unlikely that regulatory bodies will be able to treat single-cell or fermentation protein as a single class of substance. Distinctions may be made in practice between at least three groups depending on the *substrate* used: (*a*) sugars, starches and other substances from good quality agricultural materials; (*b*) hydrocarbons from petroleum or related sources; and (*c*) cellulose and other industrial wastes. Further distinctions may be necessary based on the *types* of micro-organisms used.

In the light of current information it seems likely that some regulatory bodies will insist on delaying approval for human use until there has been perhaps five years of experience with farm animals. On the other hand, ordinary yeast is normally accepted as a product of proven worth, and other types of material (for example, the microfungal proteins grown on wheat flour residues) may be able to establish a place in their own right.

The position in developing countries

How can the experience of industrialised countries be applied in low-income countries where there are often acute shortages of foods in general, and of proteins in particular? The problem can be examined from different angles – in terms of scientific knowledge, of technical processes, and of regulations and control.

Application of scientific knowledge

Over the past twenty years there has been a very considerable expansion of scientific and technical knowledge regarding proteins – and food-stuffs generally. Basic scientific facts can have universal validity and there is now much information of value that can be applied in tropical and other countries.

On the positive side, we can say with some degree of certainty that many types of agricultural produce – oilseeds and leaves, as well as cereals – can give proteins of good biological value and that mixtures of two or more plant proteins may be of special value. Thus, older phrases implying the primacy of animal proteins can in large measure be jettisoned. Our knowledge of toxic materials has also increased; for example, we know a great deal about the chemistry of aflatoxin and other substances produced by mould growth. Research warns of the dangers of toxin formation and gives guidance on the conditions to be used in harvesting or post-harvest storage of crops in order to avoid mould growth and toxin production. Investigations have also made it clear that with unsatisfactory conditions many different products can be affected, in addition to groundnuts which were first implicated. We should, therefore, avoid the error of condemning groundnuts, but rather concentrate on improved conditions, with adequate procedures for monitoring products.

Technological processes

One important question arising from North American and European work is how far developing countries should accept without question processing methods and objectives. We cannot discuss here the scale of operations, but should recognise that many of the arguments for economies of scale apply primarily to countries with a scarcity of labour and with high labour costs. Thus, to take one example, the small-scale traditional brewery exists in many towns in Africa; it is conceivable that – if sufficient effort was made – a fermentation process for protein could be worked out, based more on traditional, labour-intensive, brewery procedures than on modern capital-intensive biochemical engineering operations.

In industrialised countries research and development work on proteins has not ignored nutritional objectives, but much of the work has probably been stimulated primarily by economic and technical considerations, including (*a*) the recognition of the increasing prices of meat and animal products and the desire to obtain plant and other materials of lower cost, and (*b*) the search for materials which *functionally* can contribute to new product formulations and/or provide the basis for substitutes. It is because of (*b*) that much effort has been put into protein *isolates* made according to rigid specifications and adaptable for use in different types of products. This philosophy has led to the search for bland, colourless materials, which although of nutritional value, have to be regarded as building materials for food preparations, rather than as foods in their own right. In practice these 'unit materials' are incorporated in mixtures and may be supplemented or fortified (for example, by fats and vitamins) before use.

Inevitably the above concepts are embodied in regulations and in standards; developing countries, before adopting standards – or processing methods – are therefore entitled to ask how far these concepts are valid outside a given industrial or geographical area. If we take groundnuts as an example, a surprisingly large number of people outside North America are unaware of the fact that (apart from roasted peanuts) peanut butter is by far the simplest protein food derived from the peanut, and that it can be prepared easily on a cottage or small factory scale. In comparison with the volume of work on soya isolates, there has been only a trivial research investment in simple methods of producing peanut protein concentrates for use as a supplement in bread and other foods.

A similar point can be made about other produce. The diffusion of knowledge about the many Asian soya recipes could be – in some countries at least – more effective in the promotion of nutrition than the use of soya isolates. There are good reasons, therefore, for believing that developing countries should not be so blinded by the success of the 'isolate story', that they ignore the importance of seeking other techniques for the conversion of indigenous products. Two examples can be given; the first in relation to leaf protein, and the second to fish protein. Leaf protein concentrate may contain 25 % or more of lipid material including components with the nutritionally important carotene and poly-unsaturated acids. As a food, the concentrate contributes not only protein, but also contributes (through the lipids) to *energy* needs and (potentially) to the requirements for minor constituents. Fish meal concentrate, as normally prepared, is highly purified; on the other hand a crude fish protein mixture prepared in West Africa in 1959 contained many components including lipids; it was stable for several months in northern Ghana. The preparation tasted like fish, was readily acceptable by the local people and could be used directly in soups with the minimum of technology; however, unlike the United States' fish protein concentrates, it could not have been used as a 'neutral' protein supplement.

Legislation

Many low-income countries have no codified food law; they will almost certainly begin to incorporate aspects of the *Codex Alimentarius* in 'local' legislation or regulations. They will also wish to make use of the Guidelines on proteins issued by the PAG of the United Nations. These Guidelines have been written in practical terms and are designed to help developing countries introducing new protein products. In the interpretation of the guidelines a distinction must be made between (*a*) protein foods to be introduced as milk substitutes for babies or weaning foods for infants, and (*b*) the use of protein supplements to other food. In respect of (*a*), high nutritional and hygienic quality standards will be advocated by most parents as well as by paediatricians; for (*b*), practical circumstances dictate less rigid standards.

Food standards in the United States are high in contrast to some European countries. On the other hand, people in European countries have survived and flourished without the present complex of regulations now becoming characteristic of an urbanised western society. Developing countries may, therefore, be well advised to begin with relatively simple legislation and to avoid regulations which could hinder the supply of potentially valuable local protein and other products. The danger is not just hypothetical. To take one example, the term *enrichment* was defined in one country largely in terms of added vitamins; a legal decision was made that it could not be used for bread 'enriched' with soya and yeast extracts. Yet in developing countries it may be far cheaper and more practical to secure nutritional improvements by food *mixtures* rather than through the addition of synthetic vitamins or amino acids. A second example concerns aflatoxin and groundnuts. A national government can, with advantage, initiate an educational campaign among farmers to ensure that groundnuts are harvested and stored under suitable conditions; it may establish machinery for the chemical examination of batches of groundnuts; but it may be advised to avoid legislation which could, if implemented, virtually rule out of use groundnuts (or some other potentially valuable commodity).

Decisions about legislation should be made by *local* people who are able to weigh advantages and disadvantages of a regulation in terms of the good of their own communities, and – where doubt exists – balance (for example) the hypothetical dangers in an adult population in forty years time, against the knowledge that, unless food and other conditions are improved, 50 % of their children may die before reaching school age. These decisions are difficult to make, but the very difficulty points to the necessity for a larger pool, in the developing countries, of men and women with a wider and deeper knowledge of food and nutrition science. Such specialists are required for many reasons: for food control laboratories and for research and development work; for teaching at different levels both within the formal educational system and in adult education and extension; and not least in advising Ministries and governmental agencies on food and nutrition policy, including the implication of social, economic and legislative changes on food policy.

References

National and International Regulations

(1) *Codex Alimentarius.* Food Policy and Nutrition Division, Food and Agriculture Organisation of United Nations, 00100 Rome, Italy.

See also Nutrition News Letter, notes and reviews on Food Legislation. FAO.

(2) Protein Advisory Group of United Nations, System Room A606, United Nations, New York 10017, USA.

PAG Bulletin, duplicated Reports and Guidelines (see below).

(3) From different countries, e.g.,

Food and Drug Administration, United States Department of Agriculture, Washington DC, USA.
(documents published in the Federal Register)

Food Standards Committee and Food Additives and Contaminants Committee, Ministry of Agriculture, Fisheries and Food, Great Westminster House, Horseferry Road, London SW1.
(reports published through HMSO)

Protein Advisory Group of the United Nations

Some Policy Statements

No. 19, Maintenance and improvement of nutritional quality of protein foods
No. 23, PAG recommendations for the promotion of processed protein foods for vulnerable groups

Some Guidelines

No. 2, Preparing food quality groundnut flour
No. 4, Preparation of edible cottonseed protein concentrate
No. 5, Edible, heat-processed soy grits and flour
No. 6, Preclinical testing of novel sources of protein
No. 7, Human testing of supplementary food mixtures
No. 8, Protein-rich mixtures for use as supplementary foods
No. 9, Fish protein concentrates for human consumption
No. 10, Marketing of protein-rich foods in developing countries
No. 11, Sanitary production and use of dry protein foods
No. 12, Production of single cell protein for human consumption
No. 13, Preparation of milk substitutes of vegetable origin and toned milk containing vegetable protein
No. 14, Preparation of defatted edible sesame flour

26. Acceptance of novel foods by the consumer

R. P. DEVADAS

Development of novel foods

Attempts have been made in recent years to produce novel foods in the form of vegetable protein mixtures such as the Indian MPF (Multi-Purpose Food) (Subrahmanyam *et al.*, 1957), Incaparina (Bressani & Elias, 1966), 'High Protein Foods' and 'Complete Protein Foods' (Anderfelt, 1969), and leaf protein (Pirie, 1971). However, the use of novel foods has been restricted both by lack of attractive products, and lack of education with regard to their acceptability (FAO, 1957; Burgess & Dean, 1962). Therefore, dissemination of information to the consumer about novel foods and stepping up their production and acceptability are immediate needs.

Acceptability of foods

Any new product, however good it may be from the economic and nutritional standpoints, needs to be accepted by the individual consumer and the community at large (Devadas, 1967). Acceptance can be defined as an experience or feature of experience characterised by a positive attitude; that is, preference or liking for a specific food item (Amerine *et al.*, 1965). Among the several factors which determine the acceptability of novel foods, palatability is foremost.

Palatability

With some foods, there is a direct relationship between palatability and nutritional value. On the other hand, foods which have high nutritional value, may have low palatability (Kamalanathan *et al.*, 1969). Conversely, foods of superior palatability are not always high in nutritive value. Acceptability and palatability are conditioned by three aspects. These are the sensory properties or stimuli coming from the food, the attitudes of the eater towards the food and the consumer's interests.

Sensory properties

The sense of sight plays a considerable role in the acceptability of foods. Appearance has great influence since visual properties facilitate the selection of food from a variety of choices placed before the eye; it thus succeeds or fails to stimulate appetite. The appeal to the eye is through colour, form, size and shape, which attract the consumer to taste and eat a novel food.

Taste is the sensation realised when foods pass over the tongue and soft palate. It may be sweet, bitter, acid, or salt. It plays a dominant role in making foods acceptable or otherwise as a food must taste good to be accepted.

Flavour is that quality imparted at the nose by taste and smell together. It stimulates the urge to taste and eat food and is responsible for the overall preference and continued use of certain food products by the consumer.

Texture, which is a composite property, is another important attribute in the acceptability of a new food. The feel of the food, its crispness or softness and its consistency – fluid, semi-solid or solid – are all important, especially to children.

The effect of temperature is important because of its own effect on the palate, the consumer's associations and expectations of certain temperatures with certain foods, and the change in sensitivity to the primary tastes with variations in temperature.

Attitude of the eater

Several factors determine the attitudes, likes and dislikes of people for particular foods. Early childhood experiences with food are powerful influences in this direction. Often food preparations are rejected because they are new, strange and never seen before. Some of the factors that influence children's attitudes towards foods are described below.

What the child eats depends entirely on the mother's attitudes and what the mother chooses depends chiefly on the age of the child. The mother's choice, especially for the infant and young child, is influenced by what she believes is good for the health of the child. As the child gets older, the mother's choice becomes less important.

Emotional factors play an important role in the choice of food. One of the reasons for giving children particular foods, is that it is part of the mores of the society. Most mothers are influenced by the foods which the child especially likes or dislikes. His preference for a particular food may be due to the fact that it is liked by those whom he admires. The converse also occurs, in that he may not want to eat some foods, because they are eaten by people whom he dislikes.

Environmental factors at home and school also determine whether or not children eat a particular food. For example, the use of papaya in school lunch programmes is dependent mostly on the presence or absence of a school garden. In families with small incomes, the scope for the child to choose foods may be limited.

Cultural factors determine what people regard as good food, and how the total family meal is distributed. They explain why man anywhere in the world does not eat the full range of nutritious foods available to him. Food taboos are common even among the 'educated', and one group's food is not necessarily acceptable to people in another community.

Consumer's interests

From the consumer's point of view there are a number of practical features that will influence the acceptability of novel foods.

First, the cost must be comparable to, or preferably less than, the corresponding foods already on the market. The price is in turn determined by the cost of raw materials, processing and distribution.

The food must also be available at convenient places of purchase. What is available depends on factors such as climate, soil, competitive demands on land, storage facilities, quantity that can be stored, shelf-life, transportation to distant villages, and demand.

Convenience in preparation is a valuable demand factor, and the novel food must lend itself easily to incorporation into known recipes without many additional preliminaries and manipulative techniques in cooking. This is true in India even among the low income groups.

Rejection often occurs because of wrong notions about its digestibility. Therefore, the digestibility of the new foods must be established and the consumers convinced about its digestive qualities.

If it is to be used for community feeding programmes in the rural areas the food should have long storage life, without refrigeration, under the conditions of heat and humidity prevailing in several developing countries, most of which are in the tropical region and have problems of transportation and distribution. However, for retail consumption in urban areas, storage is not an important problem, and storage problems do not arise with many foods produced in villages and consumed there. Packaging must preserve the product while looking attractive and must also be at a cost suitable to its contents.

Incorporation of a novel food into the standard items of existing diets offers the best prospect for acceptance. The consumer must be approached, induced and convinced to buy the product for her family, and adopt it as part of the family diet pattern. Ideally it should be rich in energy, proteins and other nutrients so that one or two servings of the food in which it is incorporated can fill the nutritional gaps that exist in the diet. In India, the food intakes of pre-school children show a gap of 1.3 to 2.1 MJ (ICMR, 1971). These can be provided by 75 to 100 g of cereal–legume mixtures. However, such mixtures cannot meet the requirements of vitamins and minerals most of which are also deficient in the diets, and nutrient-rich novel foods may help to solve this problem.

Steps to be taken to make novel foods acceptable

The steps which need to be taken for making the novel food acceptable are thus:

(*a*) Obtaining knowledge of the existing dietary patterns of the consumers.

(*b*) Locating the type of food preparations in which the novel food can be incorporated.

(*c*) Developing and standardising suitable recipes, utilising the novel foods.

(*d*) Testing the standardised new recipes and ensuring their acceptability.

(*e*) Introducing the accepted preparations into homes and community feeding programmes.

(*f*) Imparting nutrition education to parents and others on the benefits of the novel foods.

(*g*) Popularising the novel foods through the press, radio, women's clubs and other media.

Acceptability of foods in nutrition intervention programmes

Very few attempts have been made to determine what factors can make the food components used in community feeding programmes more attractive from the recipients' point of view. To sell 'nutrition' *per se* to families takes considerable time and effort. Taste attributes are important considerations in determining the food commodities used in a sponsored feeding programme and in ensuring that the food will be eaten (PFA, 1969). The experience of a 'Take Home Food' project demonstrated that rural consumers were highly sensitive to the sensory attributes of a food, especially when distributed free of cost (Devadas & Murthy, 1972). A 'Take Home' type of distribution system, unlike 'On the Spot' feeding, allows ample time for experimentation with and scrutiny of the novel food by the mother before she gives it to a young child. The mother, father, older siblings and even grandparents also get a chance to examine the food for its taste before the child is given the 'test' food.

An innovation like a novel food product has to pass through several stages. In order to reach the child, the product must first meet parental approval, no matter how it is distributed (Hammonds & Wunderle, 1972). This recognition has resulted in manufacturers in the developing countries incorporating food additives in processed foods intended for children, so as to bring the food into line with adult tastes.

A trial was conducted recently to test the acceptability of an extruded food in bite-sized pieces, versus the same food in powdered form (Cantor/ATAC, 1972). The results indicated that the bite-sized pieces were preferred to the powdered form. Mothers had more well defined likes (70 % for pieces compared to 6 % for the powdered) than children (57 % liked pieces compared to 16 % who liked the powder). In their acceptability tests with CSM (Corn Soya Milk), Devadas *et al.* (1973), found adult consumers showed distinct and consistent taste and flavour preferences, while children under three years of age appeared to like all forms of the CSM and weaning mixtures prepared out of local foods, even when they were bland and unsweetened. Thus, adults considered sweetness as the overriding factor for high acceptability among infants, but the judgement of the children themselves was different. Furthermore, their studies indicated that while sweetness improved acceptability, flavour did not play an important role.

In a study on infants 4 to 7 months old, Fomon *et al.* (1970) found that modifying the taste of strained baby foods with salt, did not alter acceptability. In an extensive study of food habits conducted in Calcutta by USAID (1972), the concept of flavour was not commonly known in the low-income groups, even among urban women.

Food habits

Food habits are not ready made, but develop gradually from infancy to adulthood. Infancy and childhood are thus the best times to develop favourable attitudes towards new foods. In most cultures, the diets of young children and of pregnant and nursing women, who are the most vulnerable groups from the nutritional stand point, are based largely on one or two cereal staples. The staple food is bound up in a variety of ceremonies and people think of food in terms of their particular staple cereal; e.g., corn or maize (Central America), wheat (Europe, USA, UK) or rice (Asia). A cultural 'staple cereal' is a food that is not only the community's main source of energy, but one that has a tremendous emotional, historical, mystical and religious hold. In India and other developing countries, the more expensive and tastier items of food such as eggs and meat are the foods considered 'not good' for young children and pregnant and nursing mothers, so are only eaten by men.

Introducing novel food preparations to the consumer

Several factors favour the acceptance of new foods by consumers in developing economies. Because many householders work outside the home, there is a trend towards greater consumption of processed or ready-to-eat foods. Greater emphasis is now placed on new trade and brand names, and on attractive containers. The practices of the elite in society are becoming better known, and conventional foods are scarce.

However, it is being recognised that sophisticated processed foods are expensive and hard to store and distribute in rural areas. Therefore indigenous multimixes based on local foods appear to be the answer. Cereal–legume mixtures are the most common low cost items available in the community for formulating the mixes (Gopalan, 1970). Utilising such mixes, weaning foods just as nutritious as the most sophisticated formulas, could be developed in almost any culture (Milner, 1969).

Scientifically guided research in developing suitable mixes, formulas and recipes is being conducted in many parts of the world. Some of the new food products developed out of local cereals, pulses and oilseeds are Incaparina in Guatemala and Colombia, Fortifex in Brazil, MPF, Balahar and Lactone in India (Parpia, 1969), Superamine in Algeria, CSM in the USA, PKFM in Nigeria and Sekmama in Turkey (Manocha, 1972). Novel foods can be incorporated in legume–staple-cereal mixtures and their efficacy as supplements can be demonstrated to the mothers. Nutrition education activities should be built around those efforts without damage to people's cultural concepts.

Some governments are able to provide capital for processing and marketing new and novel foods. Consumer education plus government support can facilitate their acceptance. Emphasis should be placed on appeal through taste and texture and the packaging design of the novel foods.

In any community, some people are more predisposed than others towards accepting novel food preparations. Young boys and girls while waiting for marriage, expectant mothers who are eagerly awaiting the birth of the baby, children who are eager to grow up, labourers who are anxious to sustain their stamina and working capacity, are the groups to be approached first to try novel foods. Securing the confidence of the public through personal example by consuming the preparations advocated, home visits and demonstrations are important for achieving success.

Conclusions

Cost, availability, ease of preparation, keeping quality and packaging are some of the criteria which determine the acceptability of novel foods by the consumer. Staple cereals are not only energy sources but also have a tremendous emotional, historical and religious hold on the community. Scientifically guided research in developing suitable cereal–legume mixes, recipes and formulas, has a great potential for future nutrition intervention programmes. Nutrition education activities should be built around all those efforts.

References

Amerine, M. A., Pangborn, R. M. & Roessler, E. B. (1965). *Principles o, sensory evaluation of food.* Academic Press, New York and London.

Anderfelt, L. (1969). The role and possibilities of the food industry in influencing dietary habits. *Symposia of Swedish Nutrition Foundation,* **7,** 107–12.

Bressani, R. & Elias, L. G. (1966). All vegetable protein mixtures for human feeding – the development of INCAP vegetable mixture based on soybean flour. *Journal of Food Science,* **31,** 626–31.

Burgess, A. & Dean, R. E. A. (1962). *Food habits and malnutrition,* pp. 106–7. Tavistock Publications, London.

Cantor/ATAC (1972). Seminar on Tamil Nadu Nutrition Project, Madras (unpublished).

Devadas, R. P. (1967). Acceptability of novel proteins. *Journal of Nutrition and Dietetics,* **4,** 2.

Devadas, R. P. & Murthy, N. K. (1972). Seminar on Tamil Nadu Nutrition Project. Cantor/ATAC, Madras (unpublished).

Devadas, R. P., Roshan, B. & Murthy, N. K. (1973). Evaluation of a weaning mixture based on local foods. Report submitted to the University of Madras. (Unpublished.)

FAO (1957). FAO Nutrition Meeting Report Series 14 – Report of the Nutrition Committee for South and East Asia, Fourth Meeting. FAO, Rome.

Fomon, S. J., Thomas, C. N. & Filer, L. J. (1970). Acceptance of unsalted strained foods by normal infants. *Journal of Pediatrics*, **76**, 242.

Gopalan, C. (1970). Approaches to feeding programmes. In: *Nutritional Feeding in the Fourth Plan* (ed. T. S. Avinashilingam), p. 48. Government of Tamil Nadu.

Hammonds, T. M. & Wunderle, R. E. (1972). Nutrition intervention programmes from a market view point. *American Journal of Clinical Nutrition*, **25**, 421.

ICMR (1971). *Diet Atlas of India.* Nutrition Research Laboratories, Taranaka, Hyderabad, India.

Kamalanathan, G., Usha, M. S. & Devadas, R. P. (1969). Evaluation of the acceptability of some recipes with leaf protein concentrates. *Journal of Nutrition and Dietetics*, **6**, 12.

Manocha, S. M. (1972). *Malnutrition and retarded human development*, pp. 268–9. Charles Thomas, Springfield, Illinois.

Milner, M. (1969). Status of development and use of some unconventional proteins. In: *Protein-enriched cereal foods for world needs*, p. 97. The American Association of Cereal Chemists, St Paul, Minnesota.

Parpia, H. A. B. (1969). Proteins foods of India based on cereals, legumes and oil seed meals. In: *Protein-enriched cereal foods for world needs*, p. 129. The American Association of Cereal Chemists, St Paul, Minnesota.

PFA (1969). Protein foods for national development; operation marketing. Report on Workshop I, New Delhi, p. 65. Protein Foods Association, New Delhi.

Pirie, N. W. (1971). *Leaf Protein: Its agronomy, preparation, quality and use.* IBP Handbook 20. Blackwell Scientific, Oxford.

Subrahmanyan, V., Rama Rao, G., Kuppuswamy, S., Narayana Rao, M. & Swaminathan, M. (1957). Standardisation of conditions for the production of Indian Multipurpose Food. *Food Science*, **6**, 76.

USAID and Hindustan Thompson Associates Limited (1972). A study of food habits. Lalchand Roy & Co. (Pvt.) Ltd., Calcutta.

Index

Aerobacter aerogenes: composition of, in different media, 227
aflatoxins, produced in groundnut by *Aspergillus*, 106, 111, 112–13, 172, 236, 242; prevention of production of, 114–15; structure of, 114
Agaricus bisporus (cultivated mushroom), 202
aleurone zone of cereal seeds, 5
alfalfa, *see* lucerne
algae: blue-green, xvi, 33–4; green micro-, xvi–xvii, 35–7; grown (with bacteria) on sewage ponds, xvii, 37, 161
allergens, in castor seed, 97
Allium cepa (onion), *A. porrum* (leek), 30
Amaranthus cruentus, rate of protein synthesis by, 31
amino acid composition of proteins: of algae, 36; of broad bean, 117; of castor, 98; of cereals, 2, 3, 5, 6; of cottonseed, 85, 108; of fish meal, 217; of fungi, 203; of groundnut, 105, 107–8; of hair, wool, and feathers (keratins), 179–81; of meat, 149, 151, 180; of micro-organisms, 224, 225; of rapeseed, 70–1, 75; of sesame seed, 94, 95; of soybean, 51, 85, 108, 180, 217; of sunflower seed, 84, 85, 108; of wheat, 180; of yeasts, 216, 217, 228, 229
amino acids, requirements of young pig and chick for, 173
amino acids, essential: added to diets of non-ruminant domestic animals, 165, 166, 173, 174; complementary, in soybean and cereals, 51; in micro-organisms 223–5.
ammonium citrate: question of utilization of, by pigs, 165
anchoveta, Peruvian, 47, 197
animal fodder, sources of: algae and bacteria, xvii, 34, 37, 238–9; broad bean, 118; byproducts of wheat milling, 126; carob bean, 20; castor meal, 79, 97–9; chenopod seeds, 21; copra meal, 44; fish meal, xix, 191–2, 197, 198; fungal mycelium, 208; groundnut meal, 106–8, 109, 170; keratins, 181–4; leaf protein, 133; lupin seeds, 23; rapeseed meal, 72; safflower meal, 89, 92; sesame meal, 94, 95; soybean meal, 47, 93, 170; sugar cane, 158; sunflower meal, 79; yeast, xix, 176, 219, 220
anthrax, in wild animals, 151
arachin, protein of groundnut, 108
Arachis, *see* groundnut
arginase inhibitor, in sunflower seed, 85
arginine in protein: of *Euchlaena*, 24; of lupin seeds, 23
ascorbic acid, in soybean sprouts, 54
Aspergillus flavus, *A. parasiticus*, produce aflatoxins in groundnut, 106, 112, 114–15, 172
Aspergillus fumigatus: culture of, on straw, 205
Aspergillus niger, essential amino acids and nitrogen in, 225
Aspergillus sp., fermentation of soybean by, 55
Atriplex hortensis (garden orache), seeds of, 21
Avena, *see* oats

Bacillus subtilis, essential amino acids and nitrogen in, 225
bacteria: composition of cell walls of, 223, 224; grown (with algae) on sewage ponds, xvii, 37, 161; as source of protein for pigs, 166; variable composition of, 226–7; *see also individual species*
barley (*Hordeum vulgare*): nutrient value of old and new cultivars of, 1, 3, 4, 5, 6, 174; protein concentrates from, 129; world production of, 122
Basella alba, rate of protein synthesis by, 31
beans: dried, canned and frozen, 28; *see also* broad bean, *Phaseolus, etc.*
Beta vulgaris (beetroot, sugar beet), 30; leaf protein from, 135
biological value of proteins: of algae, 36, 37; of broad bean, 117; of cereals (old and new cultivars), 3; of copra, 44; of groundnut, 108; of rapeseed, 75; of safflower seed, 91; of sesame seed products, 96; of *Spirulina*, 34; of sunflower seed, 85
bison, American (*Bison bison canadensis*), dressed carcase as percentage of live-weight of, 148, 149
biuret, compared with urea as source of nitrogen for cattle, 160–1
blue-green algae, as food, xvi, 33–4
Brassica campestris, *B. juncea*, *B. napus*, 65, 67, 68, 69; *see also* rapeseed

Index

Brassica oleracea (Brussels sprouts, cabbage, savoy), 30, 31
Brassica tournefortii, 68
bread and baked goods, addition to: of quinoa flour, 21; of safflower seed flour, 93; of soybean flour, 58, 236; of sunflower seed flour, 87; of wool flour, 184, 185
broad bean (*Vicia faba*), 117–18, 287–8; leaf protein from, 135
brucellosis, in wild ruminants, 151
buckwheat, *see Fagopyrum*
buffalo, African (*Syncerus caffer*), dressed carcase as percentage of liveweight of, 148, 149

Cajanus cajan (red gram, pigeon pea), 9; breeding for improvement of, 9–12
calcium: in *Sphenostylis*, 25; in sunflower seed, 80
calves, hydrocarbon-grown yeast as food for, 219, 220
Candida utilis (torula yeast): composition of, with different growth rates, 228–9; culture of, 204, 205
canning: of fish, 191, 195, 196; of vegetables, 26, 31, 52
carbon dioxide: in alkaline lakes, 33; for culture of green algae, 35
carcinogens, aflatoxins as, 113
carob, *see Ceratoma*
carotenoids: in leaf protein concentrate, 242; in *Spirulina*, 34; in sunflower seed, 80; in vegetables, 27
carrots, yields of, 31
carthamin, dyestuff in safflower seed, 87
Carthamus, *see* safflower
cassava, as substrate for micro-organisms, 157, 208
castor seed (*Ricinus communis*), 96–7; meal from, 79, 97–9; protein isolates and hydrolysates from, 99; world production of 96–7
cattle: composition of carcases of, 148, 150; dressed carcase as percentage of liveweight of, 149; estimates of efficiency of conversion of plant protein by, to meat and milk proteins, 169; non-protein nitrogen in diets for, 157–62; world production of meat and milk protein from, 171
cell walls of micro-organisms, composition of, 223, 224
cellulase, for decortication of rapeseed, 73
cellulose, fungal breakdown of, 209, 210
Ceratoma siliqua (carob or locust bean), 20
cereal seeds, 1–7; distribution of protein in, 5–6, 7; protein concentrates from, 6, 121–9; staple, attitudes of consumers towards 249–50, 251
Chenopodiaceae, minor food seeds from, 20–2
Chenopodium pallidicaule (canihua), 20, 21
Chenopodium quinoa (quinoa), 20, 21
Chlorella, mass culture of, 35–7
chlorogenic acid, in sunflower seed, 81, 83, 84, 85, 87
chlorophyll, in flour made from grass, 184
chlorophyllase, in preparations of leaf protein, 137, 142
Cicer arietinum (Bengal gram, chick pea), 9; breeding for improvement of, 9–10
Citrullus vulgaris (melon), seeds of, 22–3
Clostridium, fermentation of castor meal by, 99
coconut (*Cocos nucifera*): protein products from, 43–6; world production of, 43
cod (*Gadus morrhua*), 188, 190
Codex Alimentarius, sponsored by FAO, WHO, and national governments, 234–5, 242
colour-sorting of groundnuts, electronic, 111
conarachin, protein of groundnut, 108–9
consumers: acceptance of novel foods by, xix–xx, 245–51; legislation for protection of, 233–5
Corchorus olitorus, rate of protein synthesis by, 31
Corchorus spp. (jute), leaf protein from, 135
corn, *see* maize
corn distillers' dried grains, corn germ meal, corn gluten feed, corn gluten meal, 122, 123
corn–soya–milk food (CSM), 122; powdered, and in bite-size pieces, 249
cottonseed (*Gossypium herbaceum*), xvii; amino acid composition of, 85, 108; composition of meal from, 81; detoxification of, 172, 237; strain of, free from gossypol, 176
crambe seed (*Crambe abyssinica*), detoxification of, 172
Cruciferae, toxic principles in seeds of, 170, 172, 176; *see also individual species*
Crustacea: protein in, 189; small (krill), 189, 199
cucurbit seeds, 22–3
Cucurbita pepo (pumpkin), seeds of, 22
Cysticercus regis, parasite in African buffalo and wildebeest, 151
cystine: in keratins, 181, 182; in proteins of broad bean, 117, of fungi, 203, 205, of groundnut, 108, of rapeseed, 71, 75, of sesame seed, 95, of soybean, 51 and of *Spirulina*, 34

day-length, for reproduction of soybean, 49
debittering: of safflower seed meal, 89; of soybeans, 56–7
decortication: of groundnuts, 106; of rapeseed, 66, 73–4; of sesame seed, 94; of sunflower seed, 80
developing countries: application of scientific knowledge and technological progress in, 240–2; food legislation in, 242–3; lack of capital in, 49
diammonium phosphate: question of utilisation of, by pigs, 165
digestibility: of algae, depends on processing methods, 36–7; of cereals (old and new cultivars), 3; of new foods, must be established and publicised, 247; of rapeseed protein, 75; of safflower seed protein, 91; of *Spirulina* protein, 34; of sunflower seed protein, 85; of yeasts grown on hydrocarbons, 216
diseases, common to man and wild ruminants, 151
Dolichos biflorus (horse gram, kulthi), 9; breeding for improvement of, 14–15
Dolichos lablab (walve), breeding for improvement of, 14
domestic animals, non-ruminant: as consumers and providers of protein, 169–76
domestic animals, ruminant, *see* ruminants
dressed carcase (killing-out percentage), as percentage of liveweight for domesticated and wild ruminants, 148–9
drying: of fish, 195, 197; of vegetables, 26, 52

early-maturing varieties: of legumes, 11, 12, 13, 15, 16, 105; multiple cropping with, 2
Echinochloa colona (millet), variation in, 5
eggs, estimates of efficiency of conversion of plant protein to protein of, 157–8, 169
eland (*Taurotragus oryx*): attempts to domesticate, 152–3; dressed carcase as percentage of liveweight of, 149; percentage of fat in carcase of, 148–9
elasmobranch fish: fats of, 188; world catch of, 187
Eleusine coracana (millet), variation in, 3
embryo/endosperm ratio, in cereals, 5
enzymes, inactivation of, in processing: of leaf protein, 137; of rapeseed, 68, 73, 74; of soy milk, 53
Eragostis teff (teff), breeding to improve, 7
erucic acid, in rapeseed oil, 68
Escherichia coli, essential amino acids and nitrogen in, 225
Euchlaena mexicana (teosinte), 24
Euphausia superba, in krill, 189

Fagopyrum emarginatum, *F. esculentum*, *F. tataricum*, 19–20
Fagopyrum sagittatum (buckwheat), breeding to improve, 7
fat: in carcases of red deer, cattle, and sheep, 148–9, 150; in fish, 188; in legume seeds, 23, 24; microbial, 204; in wheat concentrate, 127; in yeast grown on hydrocarbons, 216
fatty acids: in fats of ruminants, 151; in fats of teleost fish, 188; in safflower oil, 87; in *Spirulina*, 34
favism, 118, 238
feathers: amino acid composition of, 179, 180; processing of, 181–4; world production of protein in, 179
fermented foods, 208; from *Parkia*, 24; from rapeseed, 73; from soybean, xvii, 54–6
fertiliser: castor meal as, 97; for fish ponds, 193; for groundnut, 105; groundnut meal as, 109; 'whey' from leaf protein as, 138; *see also* nitrogen fertiliser
fish, xviii–xix, 187–9; increasing the consumption of, 198–9, 228; methods of utilising, 191–2, 194–7, 242; protein concentrate from, 197; world catch of, 189–90
fish farming, 192–4; hydrocarbon-grown yeast as food in, 220
fish meal: amino acid composition of, 217; essential amino acids and nitrogen in, 225; preparation of, 198; proportion of catch converted to, 190, 191–2, 196, 197; as source of lysine and methionine, 174, 175
foliar fertilisation, 2
Food and Agriculture Organisation of the United Nations (FAO), 34, 50, 234, 235, 238
freeze-drying, of fish, 197
freezing: of fish, 191, 194, 195–6; of vegetables, 28, 31, 52
fungi, 201–3; industrial production of mycelium of, 201, 203–9
Fusarium, culture of mycelium of, 204–5, 206, 209
Fusarium udum (wilt of *Cajanus*), breeding for resistance to, 10

globulins, of *Vicia* and *Pisum*, 117
gluconapin, glucosinolate in rapeseed, 69
glucosinolates, toxic constituents of rapeseed, 66, 68, 71; in different rapeseed products, 69, 75; methods of removing, 72–3; in wastes from rapeseed protein plants, 77
Glycine max, *see* soybean

Index

goat, dressed carcase as percentage of liveweight of, 149
Gossypium, *see* cottonseed
gossypol, in cottonseed, xvii, 172, 237; strain of cotton free from, 176
grass, flour prepared from, 184
'green revolution', in productivity of cereals, 1
groundnut (peanut: *Arachis hypogaea*), 105; composition of, 107; cultivation of, 105–6; formation of aflatoxins in, 106, 111, 112–15, 172, 236, 242; meal from, as fodder, 170; nutritive value of, 86, 106–9, 174; processing of, 109–11, 113; protein isolates from, 110, 111–12, 241; as source of lysine and methionine, 174, 175; world production of, 107; *see also* peanut butter
growth rate, and composition of micro-organisms, 226–9
Guizotia abyssinica (nigerseed), 23–4

haemagglutinins, in broad bean, 118
hair (cattle and pig): amino acid composition of, 179, 180; processing of, 181, 182, 184
Helianthus, *see* sunflower seed
herring (*Clupea harengus*), 188
Hibiscus esculentus, rate of protein synthesis by, 31
histidine, low in keratins, 180, 181, 183
hoofs, digestible after grinding, 181, 182
Hordeum, *see* barley
horse, dressed carcase as percentage of liveweight of, 149
hydrocarbons, yeasts grown on, 215–21, 238, 239
hydrolysates: of feathers, 181; partial, of castor meal, 99; of sesame meal, 96
2-hydroxy-arctiin, cathartic principle in safflower seed, 90

impala (*Aepyceros melampus*): dressed carcase as percentage of liveweight of, 194; percentage of fat in carcase of, 149
insects, losses of protein caused by, 7
International Standards Organisation, 234
Ipomoea aquatica, rate of protein synthesis by, 31
irrigation, of safflower, 88
isoleucine, in proteins: of cereals (old and new cultivars), 3; of safflower seed, 90, 91; of sesame seed, 94, 95; of sunflower seed, 84, 85
isothiocyanates, in rapeseed oil, 66, 71

keratinase, in processing of keratins, 182
keratins: amino acid composition of, 179–81; conversion of, to digestible form, by chemical processing, 182–4, and by grinding, 181–2; as human food, 184–5
Kerstingiella geocarpa (Hausa groundnut), 23
Kochia scoparia (burning bush, summer cypress), seeds of, 21
krill, possible utilisation of, 189

Lactobacillus casei, composition of cell walls of, 223, 224
Lactuca sativa (lettuce), 30
lamprey, vitamin A in, 188
lathyrism, 17, 18
Lathyrus sativus (khesari), 9; breeding for improvement of, 17–18
leaf protein, 133–4, 237; crops used for, 134–5; industrial production of, 141–2; lipids included in concentrates of, 241–2; nutritive value of, 138, 176; processing of, 135–8
legislation on food, 233–5; in developing countries, 242–3
legumes, improved strains of, 1, 9–18
Lens esculentus (lentil), 9; breeding for improvement of, 16–17
Lentinus edodes (shiitake), cultivated fungus, 202
leucine, high in proteins of *Euchlaena* and *Parkia*, 24
lignan glucosides, bitter and cathartic principles in safflower seed, 89–90
lipoxidase, in soybean, 51, 52
liver, aflatoxins and, 113
liver oils of fish, 188
lucerne (alfalfa: *Medicago sativa*): essential amino acids and nitrogen in meal from, 225; leaf protein from, 135, 137, 237; industrial production of leaf protein from, 141–2; nitrogenous constituents of juice of, after extraction, 167
Lupinus luteus, seeds of, 23
Lypro, lipid–protein isolate from groundnut, 110, 111–12
lysine: in cereal proteins (old and new cultivars), 2, 4, 5, 6, 7, 47, 174; cost of adding, to pig and chick diets, 174, 175; high in proteins of broad bean, 117, chenopod seeds, 21, legumes (except *Parkia*), 22, 23, 24, 25, micro-organisms, 225, ribosomes, 228, 229, and wheat concentrate, 127; low in keratins, 179, 180, 181, and in proteins of castor, 97, *Euchlaena*, 24, *Guizotia*, 24, maize, 122, safflower seed, 90–1, 92, sesame seed, 94, 95, and sunflower seed, 84, 85; more in soybean than in most plant proteins, 51; variable in proteins of coconut, 44, groundnut, 108, 174, and rapeseed, 71, 174

magnesium: loss of, from chlorophyll, in preparation of leaf protein, 137, 142
maize (*Zea mays*): embryo/endosperm ratio in, 5; multiple cropping with, 2; nutrient content of old and new cultivars of, 1, 3, 4, 5, 6, 174; protein concentrates from, 121–2; *Spirulina* as complement to, 33; world production of, 122
manatee, xviii
1-matairesinol-mono-β-D-glucoside, bitter principle in safflower seed, 90
meat: amino acid composition of, 149, 151, 180; essential amino acids and nitrogen in, 225; percentage of bone-free first and second quality of, from red deer, cattle, and sheep, 149, 150
mechanical harvesting: of sunflower seed, 28, 31; of vegetables, 28, 31
Medicago, *see* lucerne
methionine: in cereal proteins (old and new cultivars), 3, 6, 7, 47; cost of adding, to pig and chick diets, 174, 175; in fungal proteins, 203, 205; high in proteins of sesame seed, 93, 94, 95; higher in rapeseed than in soybean, 71, 75; in leaf protein, 138; in legume proteins, 11, 12, 13, 14, 23; low in keratins, 179, 180, and in proteins of broad bean, 117, 118, cereals, 47, coconut, 44, groundnut, 108, *Guizotia*, 24, micro-organisms, 225, safflower seed, 90, 91, and soybean, 51
micro-organisms: non-protein sources of nitrogen for, xix, 157; from sewage, xvii, 37, 161; variation in composition of, 233–9; *see also* algae, bacteria, yeasts
milk, world production of protein in, 171
'milks': coconut, 43–4, 45; soy, 52–3, 57–8
millets: improved strains of, 1, 4; *Spirulina* as complement to, 33; variation in, 3, 5
milling: of coconut, 44; effect of distribution of protein in grain on results of, 5–6
molasses: as cattle feed, 158, 159, 160; as substrate for fungal mycelium, 206, 208, and for yeast, 207
molluscs: culture of, 192–3; protein in, 189
Morchella, production of mycelium of, 203
moulds, losses of protein caused by, 7, 46
multiple cropping, 2, 12
mushrooms, culture and world production of, 201, 202–3
mustard, *see Sinapis alba*
mutagens, induction of variation by, 1
mycotoxins, 112, 209
myrosinase (thioglucoside glucohydrolase), in rapeseed, 66; inactivation of, 68, 73, 74
net protein utilisation (NPU): for cereals (old and new cultivars), 3; for fungi, 209; for rapeseed, 75; for *Spirulina*, 34
niacin, in chenopod seeds, 21
nigerseed, *see Guizotia*
nitrogen, non-protein sources of: for micro-organisms, xix; for pigs, 165–7; for ruminants, xviii, 157–62
nitrogen fertiliser, response to: of groundnut, 105; of rapeseed, 70; of rice, 2
nitrogen solubility index, of rapeseed protein concentrate, 75
nucleic acids: content of, in yeast, increases with temperature of cultures, 228; high content of, in algae, 37, in mycelium of *Fusarium*, 206, and in yeasts and bacteria, 166–7, 176, 226, 239; synthesis of, by *Aerobacter*, limited by phosphate supply, 227; *see also* RNA

oats (*Avena sativa*): distribution of protein in grain of, 6; nutrient content of, 2, 3, 4; protein concentrates from, 124–5; world production of, 122
oils: castor, 96; coconut, 43, 44, 45; flaxseed, 80; groundnut, 109; rapeseed, 66, 68; safflower seed, 87; sesame seed, 93; soybean, 47, 51, 80; sunflower seed, 79, 80
oilseeds: for feeding non-ruminant domestic animals, 170–1; standards for legislation on, 236–7; *see also individual species*
Oryza, *see* rice
oxalic acid, in hull of sesame seed, 94
β-N-oxalyl amino alanine, neurotoxin causing lathyrism, 17

packaging of novel foods, 247, 251
palatability: appearance, flavour, and texture in, 246, 248–9, 251; not necessarily related to nutritive value, 245
Panicum miliaceum, *P. miliare* (millets), variation in, 3
papain, for processing of keratins, 184
Parkia filicoidea (West African locust bean), 24
Paspalum scrobiculatum (kodo millet), variation in, 5
peanut, *see* groundnut
peanut butter, 236, 241
peas: dried, canned, and frozen, 28; *see also Pisum, etc.*
Penicillium notatum, essential amino acids and nitrogen in, 225
Pennisetum typhoides (pearl millet), protein and lysine content of old and new cultivars of, 4
pH: alkaline, tolerated by *Spirulina*, 33; in extraction of protein from sesame seed, 96;

pH (*cont.*)
and solubility profiles of sunflower seed meal and protein isolate, 83; of soil for soybean, 50
Phaseolus aconitifolius (moth bean), 9; breeding for improvement of, 13–14
Phaseolus aureus (moong bean, green gram), 9; breeding for improvement of, 12–13; multiple cropping with, 2
Phaseolus mungo (urd bean, black gram), 2, 9; breeding for improvement of, 13
Phaseolus spp., leaf protein from, 135
Phaseolus sublobatus, wild parent of Indian cultivated beans, 12
pheophorbide, produced in leaf juice when slowly heated, 137
phosphates: high content of, in fungal mycelium, 201; limitation of, restricts nucleic acid synthesis in *Aerobacter*, 227, 229
phytic acid, in rapeseed, 76
pigment in sunflower seed, *see* chlorogenic acid
pigs: amino acid requirements of, 173; cost of adding lysine and methionine to diets for, 174, 175; estimated efficiency of conversion of plant protein to meat protein by, 157–8, 169–70; hydrocarbon-grown yeasts as food for, 217, 219–20; non-protein nitrogen in nutrition of, 165–7; sensitive to glucosinolates in rapeseed meal, 72; world production of meat protein from, 171
Pisum sativum (garden pea, field pea), 9, 117; breeding for improvement of, 15–16; leaf protein from, 135
potato (*Solanum tuberosum*): leaf protein from, 135; multiple cropping with, 2
poultry: amino acid requirements of, 173; cost of adding lysine and methionine to diets for, 174, 175; estimates of efficiency of conversion of plant protein by, to eggs, 157–8, 169, 170, to meat, 169, 170, and to meat plus feathers, 179; hydrocarbon-grown yeasts as food for, 217, 219, 220; sensitive to glucosinolates in rapeseed meal, 72; *Spirulina* as food for, 34; world production of meat and egg protein from, 171
presses, for extraction of leaf protein, 136–7, 141
price: of novel foods, and acceptance, 247; of soybean (1972–3), 47; of *Spirulina*, 34
progoitrin, glucosinolate in rapeseed, 69
prolamines, in maize and sorghum, 3, 5
pronghorn antelope (*Antilocapra americana*), dressed carcase as percentage of liveweight of, 149
protease, fungal: hydrolysis of sesame meal by, 96
Protein Advisory Group of United Nations, 235, 242, 244
protein efficiency ratio (PER): for broad bean, 118; for chenopod seeds, 21; for copra, 44; for cottonseed, 85, for CSM food, 122; for nigerseed, 24; for rapeseed, 75; for safflower seed, 91; for sesame seed, 94, 95, 96; for soybean, 51, 118; for *Spirulina*, 34; for sunflower seed, 85, 86; for triticale, 7; for wheat flour (high protein), 127
protein nutritive value (compared with casein): for groundnut, safflower, sesame, sunflower, and soybean, 86; for safflower, 91
Psophocarpus tetragonolobus (winged bean), 24

quinoa, *see Chenopodium quinoa*

radish, fodder (*Raphanus sativus*), leaf protein from, 135
ramie (*Boehmeria nivea*), leaf protein from, 135
rapeseed (*Brassica napus*, etc.), xvii, 65–6, 67; composition of, 66; detoxification of, 72–3, 172; manufacture of oil and meal from 66, 68–70; nutritive value of, 70–2; protein concentrate and isolate from, 72–6; protein wastes from, 77; world production of, 65, 66
red deer (*Cervus elaphus*), 147; attempts to domesticate, 153; composition of carcase of, 148, 150; dressed carcase as percentage of liveweight of, 149; loss of meat in killing, 151
Rhizobium japonicum, inoculation of soybean seed with, 50
Rhizopus: culture of mycelium of, 204–5, 208; fermentation of soybean by, 55
Rhizopus nigricans, essential amino acids and nitrogen in, 225
ribosomal proteins: high in arginine and lysine, and low in serine, 228, 229
rice (*Oryza sativa*): distribution of protein in grain of, 5–6; nutrient content of old and new cultivars of, 1, 2, 3, 4; protein concentrates from, 122–4; world production of, 122
ricin, toxic protein of castor seed, 97
ricinine, alkaloid of castor seed, 97
ricinoleic acid, in castor oil, 96
Ricinus, *see* castor
RNA: ratio of protein to, in micro-organisms, is proportional to rate of growth, 226, 227, 228

ruminants, domestic, xvii, xviii; dressed carcase as percentage of liveweight, for domesticated and wild species of, 148–9; leaf protein residues as feed for, 133; not affected by glucosinolates in rapeseed, 72; percentage of protein in forage converted to human food by, 133; use of non-protein nitrogen by, 157–62
ruminants, wild, xvii-xviii, 147–8; cropping of populations of, 152; meat from, 148–51; possible domestication of, 152–3; possible transmission to man of diseases of, 151–2
rye (*Secale cereale*): hybrid of wheat and, 7; nutrient content of, 3; protein concentrates from, 128; world production of, 122

Saccharomyces (yeast), culture of, 205
safflower seed (*Carthamus tinctorius*), 87–8; composition of, 88–9; nutritive value of, 86, 90–3; protein concentrates and isolates from, 89–90, 93; world production of, 87, 88
Salmo irideus (rainbow trout), culture of, 193
Salsola kali (prickly saltwort), seeds of, 21
salting of fish, 191, 195
saponin, in chenopod seeds, 21
Scenedesmus acutus, mass culture of, 35–7
Secale, *see* rye
serine, low in ribosomal proteins, 228, 229
sesame seed (*Sesamum indicum*), 93–4; nutritive value of, 86, 94–5, 96; protein isolates and hydrolysates from, 95–6; world production of, 93
Setaria italica (Italian millet), variation in, 5
sewage ponds, mixture of algae and bacteria grown on, xvii, 37, 161
sheep: amino acid composition of meat from, 180; composition of carcase of, 148, 150; dressed carcase as percentage of liveweight of, 149; estimate of efficiency of conversion of plant protein to meat protein by, 169
shellfish (molluscs and crustaceans), 189; world catch of, 187
Sinapis alba (mustard): leaf protein from, 135; meal from seeds of, 70; oil from seeds of, 68
single-cell protein, 210; standards for, 238–9; variation in composition of, 223–9; *see also* algae, bacteria, yeasts
sodium sulphide, sodium thioglycollate, in processing of keratins, 182, 183
Solanum lycopersicum (tomato), 30
Solanum tuberosum, *see* potato
solubility profiles, for sunflower seed products, 82, 83, 87
solvent extraction: of groundnut, 109, 110; of oilseeds, 172; of rapeseed, 68–9; of safflower seed, 90; of sesame seed, 94; of soybean, 47, 49, 58; of sunflower seed, 79, 81, 85; of yeast grown on hydrocarbons, 216
Sorghum: distribution of protein in grain of, 6; nutrient content of old and new cultivars of, 1, 3, 4; protein concentrates from, 6, 128–9; variation in amino acid composition of protein of, 176; world production of, 122
soybean (*Glycine max*), xvii, 47; Asian recipes for using, 241; composition of, 50–1; composition of meal from, 70, 217; cultivation of, 49–50; as fodder, 47, 93, 170; nutritive value of, 86; processing of, into foods, 52–4, (fermented foods) 54–6; processing of, by new methods, with defatting, 58–9, and without defatting, 56–8; protein of, 51, 85, 86, 108, 180; protein concentrates and isolates from, 51; as source of lysine and methionine, 174, 175; world production of, 47–9
Sphenostylis stenocarpa (yam bean), 25
Spirulina (blue-green alga), 33; filtration of, 33–4; nutritive value of, 34
Sporotrichum thermophile, culture of, 209
springbok (*Antidorcas marsupialis*), dressed carcase as percentage of liveweight of, 149
squalene, in liver oil of elasmobranchs, 188
Staphylococcus aureus, essential amino acids and nitrogen in, 225
starch: culture of fungi on, 205; from sorghum, 128; from wheat, 126
sugar cane: as cattle food, after derinding, 158; as substrate for micro-organisms, 157
sulphur content, of legumes, 11, 12, 14
sulphur-containing amino acids, *see* methionine, cystine
sunflower seed (*Helianthus annuus*), 79; composition of, 80, 81; nutritive value of, 84–6; protein concentrates and isolates from, 80–4, 86–7; world production of, 80
sweetness in infant foods, attitudes of adults and infants towards, 249

taboos on foods, 247
taeniasis, in wild animals, 151
'Take Home Food' projects, 248–9
teleost fish, world catch of, 187
Telfairia occidentalis (fluted pumpkin), seeds of, 23
temperature: in processing of groundnut, 110, 111, rapeseed, 74, and safflower seed, 85; variation of composition of micro-organisms with, 226, 228

teosinte, *see Euchlaena*
threonine, in proteins: of broad bean, 117; of cereals (old and new cultivars), 3, 47; of coconut, 44; of lupin seeds, 23; of micro-organisms, 225
thyroid gland, affected by glucosinolates in rapeseed, 72
Tilapia (freshwater fish), culture of, 193, 194
tofu (soybean curd), 35
toxicological tests, on hydrocarbon-grown yeast, 216–17, 218
toxins in food, 233
trehalose, in fungi, 202
Tricholoma nudum, essential amino acids and nitrogen in, 225
Trichosporum pullulans (yeast), culture of, 204
triticale (hybrid of wheat and rye), 7; protein concentrates from, 127–8
Triticum, see wheat
trypsin inhibitors: in broad bean, 118; in groundnut, 107; in soybean, 51; in sunflower seed, 85
tryptophan: in keratins, 179, 180, 181; in proteins of broad bean, 117, of castor, 97, of cereals, 47, of *Kerstingiella*, 23, and of maize, 5, 122
tsetse fly, effects on wild ruminants of eradication programmes to control, 152
tuna, vitamin D in liver of, 188
tyrosine, in protein of lupin seeds, 23

United Nations Children's Fund (UNICEF) 235
urea: question of utilisation of, by pigs, 165–6; as source of nitrogen for cattle, 158, 159, 160
urease, in soybean, 51

vegetables, 15–16, 27–31; soybean as, 52
Vicia, see broad bean
Vigna sinensis (cowpea), 9, 22; breeding for improvement of, 15
Vigna unguiculata (cowpea), leaf protein from, 135
5-vinyl-2-oxazolidinethione, in rapeseed oil, 66, 71
vitamin A and vitamin D, in fish, 188
vitamin B group: destroyed in processing of vegetables, 27; in sunflower seed, 80; in teleost fish, 189
vitamin B_{12}, extract of rapeseed meal as substrate for micro-biological synthesis of, 77
Voandzeia subterranea (Bambarra groundnut), 19
Volvariella volvacea (padi-straw mushroom), 202

wastes: animal, as sources of nitrogen for cattle, 161; from rapeseed protein plants, 74, 77
Wesselsbron disease, in wild animals, 151
wheat (*Triticum aestivum*): amino acid composition of, 180; hybrid of rye and, 7; leaf protein from, 135, 137; multiple cropping with, 2; nutrient content of old and few cultivars of, 1, 4; protein concentrates from, 125–7, 128; world production of, 122
'whey': from leaf protein, as fertiliser, 138; from soybean, 51
wildebeest (*Connochaetes taurinus*), dressed carcase as percentage of liveweight of, 149
wool: amino acid composition of, 179, 180; processing of, 181, 182, 183; world production of protein in, 179
World Health Organisation of United Nations (WHO), 234, 235, 238

xanthophylls: in lucerne leaves, 141; in *Spirulina*, 34

yeasts: cell walls of, 223; culture of, 204, 205; essential amino acids and nitrogen in, 225; for food and fodder, xix, 166, 176, 239; grown on hydrocarbons, 215–17, 220–1; nutritive value of hydrocarbon-grown, 217–20; variable composition of, 228–9
yields of protein per hectare, 2; broad bean, 117; castor seed, 97; chenopod seeds, 21; cowpea, 15; flaxseed, 81; groundnut, 109; leaf protein, 135, 136; legumes, 11, 13, 15, 17; meat from cattle fed on sugar cane, 158; mushrooms, 202; safflower seed, 88; sesame seed, 93–4; soybean, 50, 81; *Spirulina*, 33; sunflower seed, 80, 81; vegetables, 27

Zea mays, see maize
zebu (*Bos indicus*), percentage of fat in carcase of, 148